Y0-BVN-403

Competition and Growth

Competition and Growth

Innovations and Selection in Industry Evolution

Jati K. Sengupta
University of California
Santa Barbara
USA

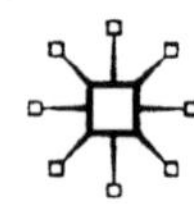

First published 2004 by
PALGRAVE MACMILLAN
Houndmills, Basingstoke, Hampshire RG21 6XS and
175 Fifth Avenue, New York, N.Y. 10010
Companies and representatives throughout the world.

PALGRAVE MACMILLAN is the global academic imprint of the Palgrave Macmillan division of St. Martin's Press, LLC and of Palgrave Macmillan Ltd. Macmillan® is a registered trademark in the United States, United Kingdom and other countries. Palgrave is a registered trademark in the European Union and other countries.

ISBN 1–4039–4164–5 hardback

This book is printed on paper suitable for recycling and made from fully managed and sustained forest sources.

A catalogue record for this book is available from the British Library.

Library of Congress Cataloging-in-Publication Data
Sengupta, Jatikumar.
Competition and growth: innovation and selection in industry evolution / Jati K. Sengupta.
p. cm.
Includes bibliographical references and index.
ISBN 1–4039–4164–5
1. Industrial organization (Economic theory) 2. Industries–Technological innovations. 3. Competition. I. Title.

HD2326.S43 2005
338—dc22 2004053142

10 9 8 7 6 5 4 3 2 1
13 12 11 10 09 08 07 06 05 04

Printed and bound in Great Britain by
Antony Rowe Ltd, Chippenham and Eastbourne.

For life's quest for the sacred

Contents

Preface

This book discusses why firms grow and decline. How does competition affect this process? Industry evolution today depends critically on innovations and R&D investments. The book analyzes the theory of Schumpeterian innovations in many forms and its impact on the selection and adjustment process in industry evolution.

Both Walrasian and non-Walrasian adjustment mechanisms in evolution are discussed here in terms of core competence and efficiency theory. The stochastic aspects of the entry and exit process and the nonparametric treatment of the R&D externalities provide some new insights. The book emphasizes the applied and empirical aspects of evolutionary dynamics and as a case study the computer industry is analyzed in some detail over the years 1985–2000 in respect of innovation efficiency, learning-by-doing and the R&D spillover effect through demand growth.

Finally, I wish to record my deepest appreciation to my two great teachers: to my Guru for his advice to lead a life dedicated to the Divine and to my father who taught me to always seek inner bliss in life.

Jati K. Sengupta

1
Competition and Growth

1.1 Introduction

The economic theory of market competition involves two stages of decision making. One is at the firm level, where each firm chooses its output optimally by maximizing profits. This yields the rule: price equals marginal cost. The second stage involves the optimal number of firms in the total industry or market, where total supply must satisfy total demand. The first stage is internal to the firm, whereas the second stage is internal to the industry. At the equilibrium price each active firm maximizes its profits and the market clears in the sense that all demand is satisfied.

How does the firm grow or decline in such an environment? If the firms have different sizes measured in terms of capital, do the small firms grow or decline in the same proportion as the large firms? The law of proportional growth implies that for given demand-growth the firms grow at a constant rate. Empirical studies however find significant deviations from the proportional growth law. For example Mansfield (1962), Jovanovic (1982) and others have empirically found that smaller firms have higher and more variable growth rates. Some of them may decline till they exit (become extinct). A recent empirical study of 182 firms in the UK manufacturing industry over the period 1980–90 by Lansbury and Mayes (1996) shows that the competitive process is empirically found to involve not just the expansion of existing firms but new entrants who challenge the incumbents often with new innovations in the form of new processes of production and new knowledge capital. This has been called 'the churning process' within and between firms, which affect the behavior of productivity and cost-efficiency. Following Schumpeter's dynamic innovation approach D'Aveni (1994) has characterized this state as hypercompetition. He argues that this hypercompetition explains many of the entry and exit decisions of firms in the high-tech businesses of today. Here the rival competitors may fail to survive if they are not on the leading edge of the innovation efficiency frontier.

The static model of competition assumes that once the optimal or profit maximizing output level has been determined, the firm would continue to produce this output level in perpetuity. But in reality this stationary equilibrium may not hold, especially in the high-tech industry, where technology and markets change very rapidly. The intensity of competition is most frequently felt at the selection stage when the industry selects the optimum number of firms. Firm growth and decay resulting in entry (increased market share) or exit (diminished market share) change the industry composition over the years. Baumol (1967) considered a profit-maximizing growth-equilibrium model of a firm, where the *rate of growth of output* rather than its *level* is used as a decision variable and the firm is assumed to maximize the discounted net present revenue. If R_0 is the initial net revenue and g the rate of growth to be determined, then this discounted net present value may be written as

$$\text{NPV} = \sum_{t=0}^{\infty} R_0 \left(\frac{1+g}{1+i}\right)^t = R_0 \frac{1+i}{i-g} \tag{1.1}$$

where i is the firm's cost of capital and for convergence one has to assume $g < i$. But the expansion has a cost $C(g)$ which is assumed to be strictly convex. This cost function also assumes increasing costs of expansion due to financial constraints of outside financing and the internal limits of organizational capacity within the firm. In equilibrium the marginal expansion cost $C'(g)$ equals $(1+i)/(i-g)^2$.

The optimal expansion path of a growing firm meeting demand-growth is more specifically introduced through a production function by means of an adjustment cost function. Here the firm is assumed to maximize the present value of the firm V:

$$V = \int_0^{\infty} R(t)\mathrm{e}^{-rt}\,\mathrm{d}t$$

subject to

$$\begin{aligned} y(t) &= F(k(t), l(t)) - C(\dot{k}(t)) \\ R(t) &= p(t)y(t) - w(t)l(t) - h(t)I(t) \\ \dot{k}(t) &= I(t) - uk(t), \quad u \geq 0 \end{aligned} \tag{1.2}$$

where the prices $p(t), w(t), h(t)$ are determined outside the firm's control and r is the rate of interest in the competitive capital market at which the firm can borrow or lend. Here adjustment costs $C(\dot{k}(t))$ is assumed to be separable from the production function. Other forms of adjustment costs have also been considered in the literature.

A third approach to the dynamic evolution of firms is based on models of industry growth emphasizing the concept of evolutionary efficiency. This approach to economic theory originally initiated by Downie (1958) has been developed by Metcalfe (1994) and recently extended in some detail by Mazzucato (2000) in terms of the differential advantages in innovation held by large firms in industries that are highly capital-intensive and concentrated. This evolutionary theory of competition has borrowed several key ideas from Ronald Fisher (1930) and his competitive fitness model of growth of biological species and Nelson and Winter (1982) and Dosi (1984) with their ideas of endogenous economic change involving evolutionary competition among firms. This approach considers two basic economic aspects of dynamic competition, when the overall demand is growing at a certain positive rate. One is that the price $p(t)$ declines over time t at a rate equal to the rate of decline in average practice unit costs. Secondly, the rate of decline in average costs is assumed to be given by the analog of Fisher's fundamental theorem of fitness in natural selection processes. This theorem postulates that the rate of change of mean fitness $(\bar{h}_s)$ is proportional to the share of weighted variance $V_S(h)$, that is,

$$d\bar{h}_s/dt = -\alpha V_S(h) \tag{1.3}$$

$$V_S(h) = \sum_{i=1}^{n} s_i(h_i - \bar{h}_s)^2$$

where h_i is in terms of economic application the unit cost level of firm i and α is the common propensity to accumulate. Here the rate of growth of each firm is proportional to $p - h_i$, where p is unit price

$$g_i = u(p - h_i)$$

and for the population of profitable firms in the aggregate

$$g_s = \sum_{i=1}^{n} s_i g_i = \sum_{i=1}^{n} s_i u(p - \bar{h}_s)$$

with s_i is the market share and capacity share of firm i and $\bar{h}_s = \Sigma s_i h_i$ is the population average of unit costs. In Metcalfe's economic reformulation of the Fisherian fitness principle, the price p declines over time at a rate equal to the rate of decline in average best practice unit costs and the rate of decline in average unit costs $(-\mathrm{d}\bar{h}_s/\mathrm{d}t)$ is proportional to the weighted variance $V_s(h)$ of unit costs given in (1.3). Clearly this dynamic model characterizes the evolution of efficient firms, which attempt successfully to adopt the strategy to reduce minimal unit costs over time. The following dynamic features are emphasized here, which have important economic implications. First of all, the efficient firms tend to grow faster, the lower its unit costs compared to the industry average. Secondly, the covariance between α and h is negative, that is,

$$\mathrm{d}\bar{h}_s/\mathrm{d}t = C_s(h, g), \quad \text{where } C_s(h, g) = \Sigma s_i(h_i - \bar{h})(g_i - \bar{g})$$

is the covariance between unit costs and propensities to accumulate. This negative covariance tends to foster higher growth rates for the efficient firms. Finally, the larger firms if they are more efficient or profitable have greater access to external capital market compared to the less profitable ones.

Our object here is two-fold: to attempt a critical evaluation of the theories of evolution of firms in a competitive industry and to discuss how technological changes affect the market selection processes for the evolution of industry over time.

1.2 Competition over time

Consider a simple growth model of a representative competitive firm, which responds to growing demand over time by increasing its capacity output and thereby reducing its optimal average costs. The demand function is of a Cobb–Douglas form

$$Q_t = a_t p_t^{-\alpha} y_t^{\beta}; \quad p_t = \text{price}, \; y_t = \text{income} \tag{1.4}$$

The price $p_t = \gamma_t c(g)$ is proportional to minimum average cost $(\gamma_t > 1)$, where output expansion $g = \dot{Q}/Q$ changes average cost. Profit is then

$$\pi_t = (\gamma_t - 1)c(g)Q_t \tag{1.5}$$

Hence one could easily determine the equilibrium growth paths of price and output:

$$g = \dot{Q}_t/Q_t = (\dot{a}_t/a_t) - \alpha(\dot{p}_t/p_t) + \beta\dot{y}_t/y_t \quad (1.6)$$
$$\dot{p}_t/p_t = \dot{\gamma}_t/\gamma_t - \varepsilon_{gc}\dot{g}_t/g_t; \quad \gamma > 1$$

where dot denotes time derivatives and ε_{gc} is the elasticity of minimal average cost with respect to output expansion. A negative value of cost elasticity assumes that a positive output growth yields a decline in optimal average costs. The sources of such decline may be the new innovations in technology, the scale effects of R&D expenditure and the learning by doing effects of cumulative experience.

The first equation in (1.6) specifies demand growth due to time shift, falling prices and income growth. The second equation says that price declines may occur due to cost economies from output expansion and also time shift due to technological changes if $\dot{\gamma}_t < 0$. In equilibrium output growth equals the growth in demand. Profit growth $(\dot{\pi}_t/\pi_t)$ is positive, whenever $c = c(g)$ falls due to higher g or, the rate of change $(\dot{\pi}_t)$ in mark-up is positive. This growth-oriented model has several implications.

First of all, consider the model of a Marshallian diffusion process due to Metcalfe (1994), which treats the diffusion of new technology. Schumpeter's theory of innovations in new products and technology focuses on this aspect of diffusion as an evolutionary process. Let $\dot{Q}_t = Q_t$ be the supply of a new commodity at time t for which the market demand is Q_d. The diffusion model assumes that output growth $\dot{Q}/Q$ is proportional to the profitability of new technology, subject to the constraint that the unit cost depends on the scale of production of the new technology:

$$\begin{aligned} \dot{Q}/Q &\geq \dot{Q}_d/Q_d \\ \dot{Q}_d/Q_d &= b(D(p) - Q) \end{aligned} \quad (1.7)$$

Here b is the adoption coefficient assumed to be a positive constant for simplicity and $D(p)$ is the long-run demand curve for the new commodity. If capacity output growth is in equilibrium with demand-growth and price is proportional to marginal costs $p = \gamma c(Q)$ we obtain a balanced diffusion path as a logistic model:

$$\dot{Q}/Q = (\alpha - \beta Q) \quad (1.8)$$

where

$$\alpha = b(d_0 - c_0 d_1(\gamma)), \quad \beta = 1 - c_1 d_1 \gamma$$
$$c(Q) = c_0 - c_1 Q, \quad D(p) = d_0 - d_1 p.$$

The model assumes a flexible competitive price process which adjusts capacity output growth to the long-run growth in demand. There exist here several sources of equilibrium output growth. First is the diffusion parameter b. The higher the diffusion rate of new technology, the greater the output growth. This diffusion parameter b may also be related to a stochastic process involving the growth of knowledge capital for the firm and the industry. Second, if demand rises over time, for example, economies of scale in demand due to globalization of trade and the innovator has a forward looking view, it helps stimulate capacity growth. Finally, the marginal cost which equals minimum average cost tends to decline due to knowledge spillovers across different firms and industries. Sengupta (2002) has used this spillover diffusion process to empirically assess the notion of key technologies that provide positive feedback and externalities to the rest of the system.

A second major implication of the output growth model (1.6) is that it can be easily related to Baumol's growth equilibrium model (1.2) and the adjustment cost theory model (1.3). For example, on using the price equation in the form $p = c(Q)$, output growth consistent with demand-growth can be easily computed as

$$g = (\dot{a}/a) - \alpha\gamma(\dot{c}/c) + \beta(\dot{y}/y) \tag{1.9}$$

Thus if $\dot{c}/c$ falls due to learning effects of knowledge spillover, it enhances the equilibrium rate of growth of output. Also the minimal cost function can be expanded so as to include adjustment costs due to capital expansion etc., that is, $c = c(Q, g, \dot{k}_t)$. In this case the extended cost function may be strictly convex, thus yielding the same marginal condition as in Baumol's model.

1.3 Efficiency and growth of firms

Dynamics of competition have been most intense in recent years in the new technology-based industries such as computers, telecommunications and microelectronics. The key role has been played here by the cost competence and efficiency of the successful firms and their increasing market share. At the industry level this has intensified the exit rate (or declining market share) of firms, which failed to maintain the leading edge of the dynamic cost-efficiency frontier.

Our objective in the following sections is two-fold. One is to analyze the dynamics of the market selection process in this competitive environment, where cost-efficient firms prosper and grow and the less efficient ones decline and fall. The second is to analyze the cost-efficiency of a firm in a nonparametric way based only on the observed cost and output data of firms comprising the whole industry.

Our analysis follows a two-stage approach. In the first stage each firm minimizes costs, given the market price. The estimation of the firm's cost frontier is obtained through a nonparametric approach, which has been adopted in recent years. This nonparametric approach initially developed by Farrell (1957) and later generalized in the theory of data envelopment analysis (DEA), see, for example, Charnes *et al.* (1994) and Sengupta (2000) estimates the cost frontier of a firm by a convex hull method based on the observed cost and output data of all firms. Unlike the method of least squares it does not purport to estimate an *average* cost function, that is, it attempts to estimate cost-specific (input-specific) efficiency of each firm relative to all other firms in the industry.

In the second stage the market clearing price is determined in the industry by minimizing total industry costs. The dynamics of adjustment around the industry equilibrium is then analyzed by a Walrasian process where prices rise in response to excess demand and fall in response to excess supply and the firm's output adjusts according to profitability.

1.3.1 A two-stage model

In the DEA models of efficiency analysis the cost efficiency has been separately analyzed from market efficiency. Thus Athanassopoulos and Thanassoulis (1995) have analyzed market efficiency in a two-stage approach, where the relative efficiency of an individual firm in capturing its share of the total market is analyzed by a linear programming (LP) version of the DEA model. We attempt here to generalize this method by explicitly allowing a nonparametric treatment of the two stages. In the first stage we estimate a cost frontier in a quadratic convex form for a firm. The second stage allows the market selection process to select the most efficient of the firms specified to be cost-efficient in the first stage. This method is very similar to the economic approach of Farrell (1957) and Johansen (1972). Farrell applied the convex hull method of estimation of technical efficiency without using any market prices but mentioned allocative efficiency as the industry level when the input and output prices are assumed to be determined by demand–supply equilibrium in the market. Johansen used the individual firm's production frontiers to determine the industry production frontier by maximizing total industry output under the constraints imposed by the aggregate inputs and the convex technology.

We consider a convex cost function model to illustrate the dynamic cost frontier of a growth oriented firm. The cost function is assumed for simplicity to be quadratic, for example,

$$C_h = \gamma_0 + \gamma_1 y_h + \gamma_2 y_h^2 \tag{1.10}$$

In our approach we consider first the problem of estimation of the firm-specific cost frontier. Let C_h and C_h^* be the cost function and the cost frontier (i.e., minimal cost function) of firm h, where $\varepsilon_h = C_h - C_h^* \geq 0$ indicates cost inefficiency. We assume C_h^* to be quadratic and strictly convex, for example, $C_h^* = \gamma_0 + \gamma_1 y_h + \gamma_2 y_h^2$, where y is output and the parameters $\gamma_0, \gamma_1, \gamma_2$ are all positive. For estimation of these parameters by the DEA approach we set up the LP model

$$\begin{aligned} \text{Min} \quad & \varepsilon_h = C_h - C_h^* \\ \text{s.t.} \quad & C_j \geq \gamma_0 + \gamma_1 y_j + \gamma_2 y_j^2; \quad j = 1, 2, \ldots, n \end{aligned} \tag{1.11}$$

based on n observations (C_j, y_j). Here total costs C_j include both variable and fixed costs and all firms are assumed to follow a given common technology. Timmer (1971) applied a variant of this method by minimizing the sum of absolute value of errors $\sum_{h=1}^{n} \varepsilon_h$, since his interest was in *robust* estimation of the production function. Sengupta (1990) has discussed other forms of estimation including corrected ordinary least squares and the generalized method of moments. The main advantage of this type of DEA estimation is that it is firm specific. To see this more clearly[1] one may consider the dual of the LP problem (1.11):

$$\begin{aligned} \text{Min} \quad & \sum_{j=1}^{n} \lambda_j C_j \\ \text{s.t.} \quad & \sum_{j=1}^{n} \lambda_j y_j \geq y_h; \quad \sum_{j=1}^{n} \lambda_j y_j^2 \geq y_h^2 \\ & \sum_{j} \lambda_j \geq 1; \quad \lambda_j \geq 0; \quad j = 1, 2, \ldots, n. \end{aligned} \tag{1.12}$$

This can also be written as

$$\begin{aligned} &\text{Min} \quad \theta \\ &\text{s.t.} \quad \lambda \,\varepsilon\, R \end{aligned}$$

and

$$R = \left\{ \sum_{j=1}^{n} \lambda_j C_j \leq \theta\, C_h \text{ and the constraints of (1.12)} \right\} \tag{1.13}$$

and λ is the column vector with n elements (λ_j) representing nonnegative weights of the convex combination of costs for each firm. Here θ is a scalar representing a measure of inefficiency, that is, $\theta^* = 1.0$ indicates 100% efficiency and $\theta^* < 1.0$ denotes less than full efficiency, that is, relative inefficiency.

On using the Lagrangean function

$$L = -\theta + \beta(\theta C_h - \Sigma\lambda_j C_j) + \alpha_1(\Sigma\lambda_j y_j - y_h) + \alpha_2\left(\Sigma\lambda_j y_j^2 - y_h^2\right) + \beta_0(\Sigma\lambda_j - 1)$$

and applying Kuhn–Tucker conditions with respect to θ and λ_j one obtains the cost frontier for firm j, when all the slack variables are zero

$$\beta C_h = 1, \theta \text{ free in sign}$$

and

$$C_j = \gamma_0 + \gamma_1 y_j + \gamma_2 y_j^2; \quad \gamma_i \geq 0, \ \ i = 0, 1, 2 \tag{1.14}$$

where

$$\gamma_0 = \beta_0/\beta, \quad \gamma_1 = \alpha_1/\beta \quad \text{and} \quad \gamma_2 = \alpha_2/\beta$$

Note that the condition $\partial L/\partial\theta = 0$ yields the numeraire condition. On varying h in the index set $I_n = \{1, 2, \ldots, n\}$ the cost efficiency frontier for all the firms can be determined. Note that for any firm h which is less than 100% efficient, that is, $\theta^* < 1.0$, one can adjust its cost C_j to $\theta^* C_j$ so that in terms of the adjusted cost firm h will be 100% efficient. Note that if we drop the constraint $\Sigma\lambda_j y_j^2 \geq y_h^2$ from (1.12) we obtain the linear cost frontier

$$C_h = C_h^* = \gamma_0 + \gamma_1 y_h \tag{1.15}$$

Here the constraint $\sum_j \lambda_j y_j^2 \geq y_h^2$ implies that the resulting cost function is strictly convex, which yields a unique cost-minimizing output. Thus equation (1.14) may be viewed as a semiparametric method of determining the cost frontier, whereas (1.15) is usually called the nonparametric method of estimating the cost frontier. Next we consider the second stage of the market demand process which selects among the first stage efficiency set, so that the total industry cost (TC) is minimized. But since the cost frontiers of firms are not all identical, we have to assume that firms are identified by their cost structures, where each firm is assumed to belong to one of m possible types of costs, each producing a homogeneous output. Let n_j be the number of firms of type $j = 1, 2, \ldots, m$ cost structure. We now minimize TC for the whole industry, that is,

$$\text{Min}_{\{n_j, y_j\}} \quad \text{TC} = \sum_{j=1}^{m} n_j C_j(y_j) \tag{1.16}$$

$$\text{s.t.} \quad \sum_{j=1}^{m} n_j y_j \geq D; \quad (n_j, y_j) \geq 0$$

where $C_j = C_j(y_j)$ denotes the cost frontier of firm j in terms of either (1.14) or (1). Total market demand D is assumed to be given. Clearly if $D > 0$, then we must have $n_j y_j > 0$ for some $j = 1, 2, \ldots, m$. On using p as the Lagrange multiplier for the market demand supply constraint and assuming the vector $n = (n_1, n_2, \ldots, n_m)$ to be given, total industry cost TC in (1.16) is minimized if and only if the following conditions hold for given $D > 0$,

$$MC_j(y_j) - p(n, D) \geq 0 \tag{1.17}$$

and

$$y_j[\text{MC}_j(y_j) - p(n, D)] = 0, \quad \text{for all } j$$

where $\text{MC}_j = \text{MC}_j(y_j)$ is the marginal cost frontier, $\text{MC}_j = \gamma_1 + 2\gamma_2 y_j$ and $p = p(n, D)$ may be interpreted as the market clearing price, that is, the shadow price which equates total supply $S = \Sigma n_j y_j$ to demand D. Let $\hat{y}_j = \hat{y}_j(p)$ be the optimal solution to (1.16). Here price p depends on n and D and hence $\hat{y}_j(p)$ may also be written as $\hat{y}_j(n, D)$. Then clearly it follows that $\hat{y}_j$ variables are uniquely determined given D and vector n.

Optimal total costs can then be specified as

$$\mathrm{TC} = \sum_{j=1}^{m} n_j C_j(\hat{y}_j(n, D)).$$

Now consider the Lagrangean function L

$$L = -\sum_{j=1}^{m} n_j \left(\gamma_0 + \gamma_1 y_j + \gamma_2 y_j^2\right) + p\Big(\sum_j n_j y_j - D\Big).$$

The first order conditions $\partial L/\partial n_j = 0 = \partial L/\partial y_j$ yield

$$p = \mathrm{AC}_j = \gamma_0/y_j + \gamma_1 + \gamma_2 y_j$$

and

$$p = \mathrm{MC}_j = \gamma_1 + 2\gamma_2 y_j \tag{1.18}$$

which implies that $p = \min \mathrm{AC}_j$. Since n_j is an integer, the continuity of the Lagrangean function $L = L(y_j, n_j, p)$ may not hold with respect to n_j. If we ignore this integral requirement as an approximation, then the Kuhn–Tucker theorem may be easily applied to determine the optimal values $(\hat{y}_j, \hat{n}_j, p)$ by finding the saddle point $(\hat{y}, \hat{n}, \hat{p})$ of the Lagrangean L for some nonnegative vectors $\hat{y}, \hat{n}, \hat{p}$. Since the L function is strictly concave in y for a fixed $n = (n_j)$ and it is linear in n for any fixed y and p, the sufficiency conditions are also fulfilled.

In the realistic case of integral n_j however we have to adopt a different method. Now we index the firms with a continuous parameter denoted by u, which replaces the integer index j. Then

$$\tilde{C}(u) = g(u, y(u))$$

gives the total cost function of firm j. The TC function for the whole industry becomes

$$C(s) = \int_0^s \tilde{C}(u) F(u) \mathrm{d}u \quad 0 \le s \le 1$$

in place of the sum $\Sigma n_j C_j$, where $F(u)$ denotes the number of firms of type u and the upper limit of the integral s denotes the value of the index of the "marginal firm", that is, the firm that it just pays to operate, given the desired total output. The industry's total output is given by

$$Y(s) = \int_0^s y(u)F(u)\,\mathrm{d}u.$$

Now we order the firms with respect to their minimum average cost, so that it is a rising function of the continuous indexing parameter u. The optimal s and y is then the solution of the following problem

$$\begin{aligned} \mathrm{Min}_{(s,y)\geq 0} \quad & C(s) \\ \text{s.t.} \quad & Y(s) = D > 0. \end{aligned}$$

The Lagrangean is given by

$$C(s) + p\left(D - \int_0^s y(u)F(u)\,\mathrm{d}u\right).$$

If there is a minimum, then the following relations must hold:

$$\partial C(s)/\partial y - p\int_0^s F(u)\,\mathrm{d}u \geq 0$$

and

$$\partial C(s)/\partial s - py(s)F(s) \geq 0.$$

Since

$$\partial C(s)/\partial s = \tilde{C}(s)F(s)$$

and

$$\frac{\partial C(s)}{\partial y} = \int_0^s \frac{\partial \tilde{C}}{\partial y} F(u)\,\mathrm{d}u$$

it follows that

$$\int_0^s \left(\frac{\partial \tilde{C}}{\partial y} - p\right) F(u)\,\mathrm{d}u \geq 0$$

$$\tilde{C}(s) - py(s) \geq 0. \tag{1.19}$$

By hypothesis the output of the marginal form is positive, hence the necessary conditions in (1.19) become equalities. Moreover, without loss of generality we may assume that $F(u) > 0$ for all u with $0 < u < s$. These imply that

$$\frac{\partial \tilde{C}(u)}{\partial y} = p, \quad \text{for all } u \text{ with } 0 \leq u \leq s$$

and

$$\frac{\partial \tilde{C}(s)}{\partial y(s)} = \frac{\tilde{C}(s)}{y(s)}, \quad \text{that is, MC} = \text{AC}.$$

The first condition means that for all active firms (i.e., firms with positive outputs) there must be a common marginal cost. The second condition means that the scale of operation of the "marginal firm" is such that its average cost is a minimum. These are the same conditions as in (1.17). Also, for all extra marginal firms, for all $u > s$, $y(u) = 0$.

Let us now denote by $\hat{y}$ and $\hat{n}$ the optimizing values for the total cost function $C(s)$. These may be viewed as the continuous approximation of the earlier problem with a discrete number of firms. These variables satisfy $p =$ AC and AC $=$ MC. Thus the long-run equilibrium price p is the minimum point of the AC curve, that is, the minimum efficient scale (MES). We may thus use $\hat{n}_j$ and $\hat{y}_j$ as the discrete approximation of the continuous model.

Note that the specific computation of the minimum efficient scale of output is possible here due to the quadratic cost function used here for illustration. The quadratic form of the convex function in (1.11) allows us to write the dual problem (1.12) in a linear form, as is customary in the nonparametric efficiency analysis approach of data envelopment analysis. However, log linear and other nonlinear forms can be introduced in (1.11) yielding a nonlinear DEA model. But the results in (1.19) would still hold due to the convexity of the total cost function with respect to n and y.

Several economic implications of this result may now be discussed. First of all, if the market demand function is viewed in its inverse form, that is, $p = F(D)$ where $D = \hat{Y}$, $\hat{Y}$ being the aggregate output, then

$$F(Y) \geq \text{LRAC}(\hat{y}) \quad \text{as } Y \leq \hat{Y}$$

where LR denotes the long run. But since $\text{LRAC}(y_j)$ is of the form

$$\text{LRAC}(y_j) = \frac{\gamma_0}{y_j} + \gamma_1 + \gamma_2 y_j$$

its minimum is attained at

$$\hat{y}_j = (\gamma_0/\gamma_2)^{1/2} = (\gamma_{0j}/\gamma_{2j})^{1/2}$$
$$\text{LRAC}_j(\hat{y}_j) = \alpha_1 + 2\sqrt{\gamma_0\gamma_2}.$$

Thus a dynamic process of entry (or increase in market share) or exit (or decrease in market share) can be specified as a Walrasian adjustment process

$$\mathrm{d}n_j/\mathrm{d}t = k_j(p - \text{LRAC}(\hat{y}_j))$$

and

$$\mathrm{d}p/\mathrm{d}t = b(D(p) - (\hat{Y})) \tag{1.20}$$

where k_j and b are positive parameters. The equilibrium is then given by the steady state values $p^* = \text{LRAC}(\hat{y})$ and $D(p^*) = (\hat{Y})$. This entry (exit) rule (1.20) is different from the limit pricing rule developed by Gaskins (1971) and others, in that this is determined directly from the estimate of MES of cost-efficient firms. Also different cost structures are allowed here, which implies that in terms of minimum average costs different firms can be ranked. For example if firms are ordered from the lowest to highest according to minimum average costs $\hat{c}_j$ as follows:

$$\hat{c}_{(1)} < \hat{c}_{(2)} < \cdots < \hat{c}_{(k)}, \quad k \leq n$$

then $p_{(1)} = \hat{c}_{(1)}$ would be the lowest price, whereas $p_{(k)} = \hat{c}_{(k)}$ would be the highest. Hence if due to exogenous demand shift the price p comes down to $p = p_{(1)}$, then all other firms have to exit in the long run. Likewise if demand shift raises the price to $p = p_{(k)}$ then the other firms would earn positive "*rents*" as

$$\Delta_{(j)} = p_{(j)} - p_{(1)}, \quad j = 2, 3, \ldots, k \tag{1.21}$$

Low-cost firms can produce at a lower AC than the others, because they may possess some scarce factor, such as superior technology, which is not available to others. Thus the low-cost firms may earn for some time more than normal profits, that is, excess profits. Some potential entrants, seeing the large profits made by the low-cost firms, would want to adopt the superior technology, thus wiping out the extra rent. Until this happens, the low-cost firms would enjoy positive differential rent, that is, early adopter's profit advantage. Thus in the long run the variance of $\hat{c}_{(j)}$ or of differential rent $\Delta_{(j)}$ would decline, though it may be high in the short run.

Now consider the case when each firm j has a separate cost function $C_j(y_j)$, that is, $m = 1$. The industry model then takes a simple form

$$\text{Min} \quad \sum_{j=1}^{n} C_j(y_j) \tag{1.22}$$
$$\text{s.t.} \quad \Sigma y_j \geq D; \quad y_j \geq 0; \quad j \epsilon I_n.$$

By rewriting the cost function as $C_j(y_j; k_j)$ where k_j is capital endowment, short- and long-run cases can be distinguished. The short-run case assumes k_j to be constant so that the cost function depends on output only while in the long-run case the cost function depends on output and capital inputs, which are both variable. In the long-run case the Lagrangean function can be written as

$$\begin{aligned} L &= -\sum_{j=1}^{n} C_j(y_j, k_j) + p(\Sigma y_j - D) \\ &= \sum_{j=1}^{n} [py_j - C_j(y_j, k_j)] - pD \\ &= \sum_{j=1}^{n} \pi_j - pD \end{aligned} \tag{1.23}$$

where π_j is the profit function of firm j, if p is interpreted as the market clearing price. In a competitive industry each firm is a price taker, so market price is given as $\hat{p}$. In the short-run capital inputs are also given as $\hat{k}_j$. Hence the vector $Y^* = (y_1^*, y_2^*, \ldots, y_n^*)$ is a short-run industry equilibrium (SRIE) if each firm j maximizes profit π_j with respect to y_j. Since the profit function is strictly concave, this SRIE $Y^*(\hat{K})$ exists and it is unique for every given vector $\hat{K} = (\hat{k}_1, \hat{k}_2, \ldots, \hat{k}_n)$. For the long-run industry equilibrium LRIE we have to modify the objective function as long-run profits defined as

$$W_j = \int_0^\infty \mathrm{e}^{-\rho t}[\pi_j(t) - h(u_j(t)]\,\mathrm{d}t \tag{1.24}$$

where $u(t)$ is investment defined by $\mathrm{d}k_j/\mathrm{d}t = u_j(t) - \delta_j k_j(t)$, $h(u_j(t))$ is a convex cost function and δ is the fixed rate of depreciation.

The LRIE is now defined by vectors K^*, Y^* if for each firm j,

$$\begin{aligned} &\text{(i)} \quad y_j^* \text{ maximizes } W_j \text{ for given } k_j, \\ &\text{(ii)} \quad k_j^* \text{ maximizes } W_j \text{ for given } y_j^*. \end{aligned} \tag{1.25}$$

But since investment cost function is separable, the maximization problem (1.25) is equivalent to max π_j.

Here $\hat{p} = p(\hat{y}) = p(y^*)$ and $\hat{y} = y^* = \sum_{j=1}^{n} \hat{y}_j$. Thus the industry equilibrium price p^* clears the market and given p^* each firm maximizes long-run profit with respect to y_j and k_j. Now define a *competitive industry equilibrium* by vectors Y^*, K^* and price p^* such that $D = D(p^*) = \Sigma y_j^*$ and the conditions (1.25) hold. Then one can easily prove that such a competitive industry equilibrium exists, since the cost functions are strictly convex and the profit function strictly concave; see for example, Dreze and Sheshinski (1984).

1.3.2 Evolution of firms

The evolution of firms under competition can now be analyzed by the dynamic adjustments. Two types of dynamic adjustments are implicit in the two-stage model of competitive industry equilibrium developed before. One is the path of optimal accumulation of capital by each efficient firm j, which maximizes the discounted stream of profits. This involves continuous adjustment of existing capacity (capital) by an optimal investment program so as to reduce the current cost of using exiting capacity for producing current outputs. For example, consider a quadratic cost function $h_j = g_1 u_j + g_2 u_j^2$ for each firm j, which solves the dynamic problem

$$\begin{aligned} \text{Min}_{u_j} \quad & \int_0^\infty e^{-\rho t} \left[\gamma_0 + \gamma_1 y_j + \gamma_2 y_j^2 + \beta_1 k_j + \beta_2 k_j^2 + h_j \right] dt \\ \text{s.t.} \quad & \dot{k}_j = u_j - \delta k_j; \quad \dot{k}_j = dk_j/dt \\ & k_j(0) > 0 \text{ given.} \end{aligned}$$

On using the current value Hamiltonian involving k_j and u_j:

$$H = -\beta_1 k_j - \beta_2 k_j^2 - g_1 u_j - g_2 u_j^2 + \mu(u_j - \delta k_j).$$

Pontryagin's maximum principle yields the optimality conditions

$$\dot{\mu} = (\rho + \delta)\mu + \beta_1 + 2\beta_2 k_j$$
$$u_j = (2g_2)^{-1}(\mu - g_1)$$
$$\dot{k}_j = u_j - \delta k_j.$$

On eliminating μ one obtains the pair of differential equations

$$\begin{pmatrix} \dot{u}_j \\ \dot{k}_j \end{pmatrix} = \begin{bmatrix} (\rho+\delta) & \beta_2/g_2 \\ 1 & -\delta \end{bmatrix} \begin{pmatrix} u_j \\ k_j \end{pmatrix} + \begin{pmatrix} A_1 \\ 0 \end{pmatrix} \tag{1.26}$$

where $A_1 = (\beta_1 + (\rho + \delta)g_1)/(2g_2)$.

The characteristic equation is

$$\lambda^2 - \rho\lambda - [\delta(\delta + \rho) + \beta_2/g_2]. \tag{1.27}$$

Since the product of two roots is negative and the sum positive and

$$\rho^2 + 4(\beta_2/g_2 + \delta(\rho + \delta))$$

is positive, the two roots are real, one positive and one negative. Hence there exists a saddle point equilibrium. The steady state levels are given by

$$\bar{k}_j = -(\beta_1 + g_1(\rho + \delta))/2(\beta_2 + (\rho + \delta)\delta g_2)$$
$$\bar{u}_j = \delta\bar{k}_j.$$

It is clear that as the cost coefficients g_1 or β_1 rise the steady state levels of capital and hence investments decline. The two characteristic roots imply that there is a stable manifold along which the motion of the system (1.26) is purely towards $(\bar{u}_j, \bar{k}_j)$ and an unstable manifold along which motion is exclusively away from $(\bar{u}_j, \bar{k}_j)$. By transversality conditions one may choose only the stable manifold. By using this stable manifold around the steady state equilibrium one may state the following proposition.

Proposition 1. For each vector K there exists a SRIE $Y^*(K)$, where each y_j^* maximizes $\pi_j = p^* y_j - C_j(y_j, k_j)$. There also exist a LRIE given by the pair (Y^*, K^*) where for each firm j:

(*i*) y_j^* solves $\max \pi_j = p^* y_j - C_j\left(y_j, k_j^*\right)$

and

(*ii*) k_j^* and u_j^* solve the steady state level of profits

$$\overline{\pi}_j = p^* y_j^* - C_j\left(y_j^*, k_j\right) - h(u_j).$$

The SRIE and LRIE solutions are unique.

Proof. Existence follows from the fact that the production set is convex, closed and bounded by assumption. Strict concavity of the profit function yields uniqueness. Equilibrium market price $p^* = P(y^*)$, $y^* = \sum_{j=1}^{n} y_j^*$ equalizes total demand and supply.

The adjustment of the market equilibrium may be directly shown in terms of the Walrasian process of price quantity adjustments as specified in (1.20) before, for example,

$$\dot{y} = a[p - c(y)] \tag{1.28}$$
$$\dot{p} = b[D(p) - y]$$

where $c(y)$ is the long-run minimal average cost function and a, b are positive parameters. The minimal average cost function intersects the marginal cost function at the optimal output y^* and the marginal cost is linear in output for the quadratic cost frontier; also the demand function $D(p)$ is linear in this case. Hence (1.28) can be viewed as a linearized version around the optimal output level y^*. Thus the linear differential equations for (1.28) can be analyzed in terms of the characteristic roots. For nonlinear forms of $c(y)$ and $D(p)$ we have to consider linearized versions around the optimal point y^* in order to analyze the stability of the adjustment process. The conditions for convergence to the steady state of this linearized system are once again specified by its characteristic roots. Two important cases are:

(i) each root has a negative real part; this implies that the steady state equilibrium is stable in the sense of convergence to the steady state,
(ii) two real roots, one positive and one negative; this implies a saddle point equilibrium. There is a stable manifold of convergence.

Hence one can state the result.

Proposition 2. There exists a stable manifold along which the LRIE can be reached by a Walrasian adjustment process.

Proof. Here one can apply the linearizing process to the cost $c(y)$ and demand $D(p)$ functions by taking their slopes c' and d', so that the characteristic equation can be derived. Thus by using the explicit cost functions assumed to be strictly convex, the two roots of the characteristic equation can be directly computed, for example,

$$\lambda^2 + (ac' - bD')\lambda + ab(1 - c'D') = 0 \tag{1.29a}$$

where $c' = \partial C/\partial y$, $D' = \partial D/\partial p$. Equation (1.29a) has roots with negative real parts if and only if $ac' - bD' > 0$ and $(1 - c'D) > 0$ and $ab(1 - c'D') > 0$. But with $D'(p) < 0$, $c' > 0$ implies that these conditions hold. In the second case if it holds that $c' < 1/D' < 0$, then one real root is positive and the other negative. Now the two roots are

$$\lambda = -1/2(ac' - bD') \pm 1/2[(ac' - bD')^2 - 4ab(1 - c'D']^{1/2}.$$

Since $c' < 1/D'$, there are two real solutions in λ, one positive and one negative. The point (y^*, p^*) is now a saddle point. There is a stable manifold along which the motion converges towards (y^*, p^*) and an unstable manifold along which the motion is away from (y^*, p^*). The slopes of these manifolds at (y^*, p^*) are given by the eigenvectors of the matrix in (1.29b)

$$\begin{pmatrix} \dot{y} \\ \dot{p} \end{pmatrix} = \begin{pmatrix} -a\frac{\mathrm{dc}}{\mathrm{d}y} & a \\ -b & b\frac{\mathrm{bD}}{\mathrm{d}p} \end{pmatrix} \begin{pmatrix} y \\ p \end{pmatrix} \tag{1.29b}$$

corresponding to the stable and the unstable roots respectively. From this one can readily verify that the unstable manifold has a negative slope at the point (y^*, p^*). Note that in the general nonlinear case the Lyapunov theory of stability has to be applied.

Two points are to be noted. One is that the concept of convergence used here is in the long run, when the system remains in its phase space within which certain properties hold and not that it actually converges to a point. Second, in the short run the system must be such that the initial point must be "close" to equilibrium and just right as specified by the characteristic vectors corresponding to the characteristic roots.

Some important implications of the two dynamic adjustment processes have to be briefly mentioned. First of all, if individual firms do not follow these optimal paths of capital accumulation, then they would not be consistent with LRIE. Also, firms which are cost efficient in the short run may not be so in the long run unless they use an optimal investment path. Hence there is scope for analyzing inefficiency in the long run. This aspect has been analyzed in a nonparametric framework by Sengupta (2000). Secondly, the competitive industry model developed here has a decentralization interpretation in terms of firms surviving under long-run equilibrium, see for example, Gabszewicz and Michel (1991). Novshek (1980) has shown that this type of equilibrium can be extended to include the case of Cournot equilibrium, if firm size is measured by technology, market size measured by perfectly competitive demand and if firms are small relative to the overall market and free entry conditions prevail. In such a case Cournot equilibrium exists and the aggregate output is approximately competitive. Finally, one could empirically test the consistency of the cost efficiency model (1.14) estimated by the DEA model in respect of the industry equilibrium. So long as MES levels are different for cost-efficient firms, there exists some scope for improving efficiency in the long run. This implies price changes due to entry and exit of firm in the industry with a consequent impact on individual firms through allocative efficiency.

1.4 Selection process under allocative efficiency

The selection process in industry equilibrium in competitive markets may be analyzed more directly if we assume that market clearing prices are estimated by a demand function $\hat{p} = a - by$, $y = \sum_{j=1}^{n} y_j$. In this case the industry equilibrium is directly obtained from maximizing total industry profits π, where the cost function of each firm is strictly convex and quadratic:

$$\text{Max}_{y_j} \pi = \sum [\hat{p} y_j - C_j(y_j)] \tag{1.30}$$

Since $\hat{p}$ is the market clearing price, equilibrium of market demand and supply is implicit here. The industry selection process (1.30) selects optimal outputs y_j^* and $y^* = \Sigma y_j^*$ so as to meet total demand by following the rule

$$y_j^* = A_j - B_j y^* \tag{1.31}$$

$$y^* = \left(1 + \sum_j B_j\right)^{-1} \left(\sum A_j\right)$$

where $A_j = (b + 2\gamma_{2j})^{-1}(a - \gamma_{1j})$, $B_j = (b + 2\gamma_{2j})^{-1} b$.

Let c_j^* be the optimal average cost $C_j(y_j^*)/y_j^*$, then the entry or increased market share rule can be specified as

$$\dot{y}_j = k_j(\hat{p} - c_j^*), \quad k_j > 0$$

that is, entry (market share) is positive (increasing) or negative (decreasing) according as $\hat{p}$ exceeds (falls short of) c_j^*. The price adjustment in the market can be similarly specified as

$$\dot{p} = k(D(\hat{p}) - y^*), \quad k > 0.$$

The equilibrium supply behavior specified in (1.31) implies the following comparative static consequences:

$$\partial y_j^*/\partial \gamma_{2j} < 0, \quad \partial y_j^*/\partial \gamma_{1j} < 0, \quad \partial y_j^*/\partial b < 0$$

and

$$\partial y_j^*/\partial a > 0.$$

If all cost functions are identical so that $A_j = A$ and $B_j = B$ for all j, then one obtains

$$y_j^* = A - B(1 + nB)^{-1}(nA)$$
$$y_j^* = (1 + nB)^{-1}(nA); \quad p = a - by^*.$$

This shows the impact of the number of firms in the industry on equilibrium industry output.

When firms are not alike in their cost functions but belong to a cost structure, each firm may follow one of m possible types of cost. Let n_j be the number of firms of type $j = 1, 2, \ldots, m$ cost structure. Then the allocative efficiency model (1.30) takes the form

$$\text{Max}_{n_j, y_j} \pi = \sum_{j=1}^{m} \hat{p} n_j y_j - \sum_{j=1}^{m} n_j C_j(y_j)$$

where $\hat{p} = a - b\sum_j n_j y_j; \quad a, b > 0$. This yields the equilibrium conditions

$$p_j^* = \hat{p}(y_j^*) = \left(1 - \frac{1}{|\varepsilon_p|}\right)^{-1} \text{MC}_j\left(y_j^*\right) \tag{1.32}$$
$$p_j^* = \text{AC}_j\left(y_j^*\right)\left(1 - \left|\varepsilon_{n_j}\right|\right)^{-1}$$

where ε_p and ε_{n_j} are the price elasticity of demand and size elasticity of price respectively, that is,

$$\varepsilon_{n_j} = (\partial p/p)/(\partial n_j/n_j), \ \ \varepsilon_p = (\partial y_j/y_j)/(\partial p/p).$$

When $|\varepsilon_p|$ tends to infinity and ε_{n_j} tends to zero then we obtain the earlier result (1.19), that is,

$$p = \mathrm{MC}_j(y_j^*) = \mathrm{AC}_j(y_j^*), \quad y_j^* > 0.$$

Thus we can state the result:

Proposition 3. There exists an industry equilibrium specified by the pair of vectors (n^*, y^*), where $n^* = (n_1, n_2, \ldots, n_m)$ and $Y^* = (y_1^*, y_2^*, \ldots, y_n^*)$ which satisfy the optimality conditions (1.32). If each cost function is strictly convex and quadratic, then this equilibrium pair (n^*, y^*) is unique. Furthermore, there exists a stable manifold along which the equilibrium could be reached from a nearby nonequilibrium point.

Proof. Since the profit function is closed and bounded by its continuity with respect to (n, Y), the existence of industry equilibrium is assured. By strict concavity the equilibrium is unique. Furthermore, the Walrasian adjustment process defined by the free entry (exit) rules possesses a stable manifold due to the negative slope of the demand function and strict convexity of the cost function.

Several implications of this proposition are important in economic terms. First of all, optimality of the industry equilibrium (n^*, y^*) may be tested against the observed outputs y_j and firm sizes n_j. In cases of disequilibrium the observed values would differ from the optimum and hence the market process of adjustment through entry and exit has to be analyzed. Secondly, the impact of demand on equilibrium output and price can be directly evaluated in this framework. In particular, large demand fluctuations would tend to have some adverse reaction for the risk averse producers, for example, their optimal output would tend to be lower. Finally, market concentration measured by unequal firm sizes would affect the industry equilibrium, for example, firms with the least optimal average cost would survive longer.

1.5 Technology and market dynamics

Technology is viewed here as innovations in the Schumpeterian sense. In this framework technology may involve the introduction of new processes, new software, new forms of marketing, and even new organizational forms. Here competition has two facets: static and dynamic. The former takes technology as given, so firms compete only on costs and prices. Thus greater competition reduces prices and/or raises unit costs, thus reducing profits. In the limit some firms may have to exit due to dwindling profits. The dynamic or Schumpeterian competition however changes technology at various points of the value chain, thus challenging firms to compete in new ways. Hence the successful firms in such an industry transform their technologies so as to create new strategic assets, which bring them new streams of increased cash flows and profits.

In this framework of dynamic competition, growth efficiency (i.e., growth of output) rather than level efficiency (i.e., level of output) is more important. If we view overall economic efficiency as an escalation ladder, then in dynamic competition firms compete in racing up the ladder. The Schumpeterian concept of innovations implies racing up the escalation ladder in the following specific arenas: R&D investment in new processes, products and software, sharing in network research pool and sustaining strongholds by dynamic economies of scale. One may term this as *innovations efficiency* as the main feature of dynamic competition.

Our object in this section is two-fold: to formalize a model of dynamic innovations efficiency and its role in firm growth and decay and then to discuss its impact on the market structure. It is a stylized fact in firm-size dynamics that firms in today's business environment differ significantly in their commitment to innovation and their ability in innovation efficiency. Innovation in products, blueprints and technological processes is largely endogenous to the firm's activities via R&D expenditures, learning by doing and sharing in network pools. This makes market structures endogenous to the innovation process.

Once the dynamic view of competition is used as a framework for analyzing firm growth and decay, one has to analyze the mechanism by which firm size affects innovation and vice versa. Reasons why large size may be advantageous for innovations efficiency comprise the following: capital market imperfections which give preference to larger firms due to their large internal retained earnings and strongholds in strategic assets, higher returns from R&D investment, and greater complementarities between manufacturing, marketing and financial planning. Thus Schumpeter (1942) emphasized very strongly that the large-scale firm has come to be the most powerful engine of economic progress. In this respect the static framework of perfect competition has no title to being set up as a model of ideal efficiency.

1.5.1 Growth efficiency frontier

Growth and decay of firms in dynamic competition are essentially driven by efficiency. Unlike the concepts of technical and allocative efficiency of static competition, dynamic efficiency essentially comprises innovations efficiency as defined in a Schumpeterian framework. This may be easily related to the replicator dynamics in evolutionary biology:

$$\dot{x}_i = Ax_i(E_i - \overline{E}), \quad \overline{E} = \sum_{i=1}^{n} x_i E_i \tag{1.33}$$

Here x_i represents the proportion of species i in a population of interacting species, E_i is its 'reproductive fitness' and $\overline{E}$ is the 'mean' fitness. In dynamic competition x_i may represent market share of firm i, E_i its minimal average cost and $\overline{E}$ the weighted average cost in the industry. Fisher's fundamental theorem in replicator dynamics states

$$\frac{\mathrm{d}}{\mathrm{d}t}(\overline{E}_x) = -\alpha V_x, \quad \alpha > 0 \tag{1.34}$$

that is, the rate of change of mean fitness (industry efficiency level) is proportional to the variance of fitness characteristics in the population. Following our discussion of this theorem in (1.3) before, in the context of evolutionary economic growth we may easily apply this type of reasoning to explain the relative decay of firms rather than the growth. Those firms which cannot sustain dynamic efficiency over time feel the intense pressure to exit. On the other hand, those firms escalating on the dynamic efficiency ladder tend to gain footholds and grow in market share.

To be more specific, consider the growth efficiency frontier for firms which are growing over time under dynamic competition. To fix ideas let y denote the output or sales of a firm with x_1, x_2 as the two inputs: capital and labor defining a production function

$$Y = \beta_{0t} + \beta_1 X_1 + \beta_2 X_2 \tag{1.35}$$

where $Y = \ln y$ and $X_i = \ln x_i, i = 1, 2$. Now consider a firm on the technical efficiency frontier. If it has cost-efficiency in the sense of economies of scale, that is, $\beta_1 + \beta_2 > 1$, it may be difficult for a new entrant to offer competitive prices because the incumbent's higher cumulative output reduces its unit cost (e.g., learning curve effects). On taking the time derivative of equation (1.35) one can easily derive the growth equation

$$\dot{y}/y = \dot{\beta}_{0t}/\beta_{0t} + \beta_1 \dot{x}_1/x_1 + \beta_2 \dot{x}_2/x_2 \tag{1.36}$$

where $\dot{\beta}_{0t}/\beta_{0t}$ represents total factor productivity growth or Solow-type technological progress. Recently Hall (1990) has developed a modified

Solow residual by incorporating market power measured by a mark-up ratio μ and the increasing returns to scale index $\gamma = \beta_1 + \beta_2$ as follows:

$$\dot{\beta}_{0t}/\beta_{0t} = \dot{y}/y - \beta_1(\dot{x}_1/x_1) - (1-\beta_1)(\dot{x}_2/x_2) - (\mu - 1)\beta_1\left(\frac{\dot{x}_2}{x_2} - \frac{\dot{x}_1}{x_1}\right)$$
$$- (\gamma - 1)(\dot{x}_1/x_1)$$

Clearly the innovations efficiency may augment the values of both μ and γ above unity. Also, the capital input x_1 in the production function (1.35) comprises both physical and human capital, where the latter in the form of knowledge and skill and their diffusion across other firms tend to grow exponentially without the limits imposed by diminishing returns to scale. Lucas (1993) has used this aspect to emphasize the cumulative process of growth in output in a macrodynamic framework.

We now consider an optimal expansion path for a growth-oriented firm under dynamic competition. Assume the industry to be composed of N firms, where each firm has m inputs x_{ij} and a single output y_j, such that the first m_1 inputs are current and the rest capital inputs. The growth of inputs and output are denoted by $g_{ij} = \Delta x_{ij}/x_{ij}$ and $z_j = \Delta y_j/y_j$. Let the unit cost of input growth be denoted by $c_i = c_i(g_i)$ where $g_i = \Delta x_i/x_i$. Then the optimal or efficient growth path of a reference firm h can be specified by the optimization model

$$\begin{aligned} \text{Min} \quad & C = \sum_{i=1}^{m} c_i g_i \\ \text{s.t.} \quad & \sum_{j=1}^{N} g_{ij}\lambda_j \le g_i, \quad i = 1, 2, \ldots, m \\ & \sum_{j=1}^{N} z_j\lambda_j \ge z_h; \quad \Sigma\lambda_j = 1; \quad \gamma_j \ge 0. \end{aligned} \tag{1.37}$$

If the unit cost of expansion c_i is constant, then this yields a LP model. If the firm h is growth-efficient, then it is on the dynamic production frontier

$$z_h = (\beta_0^*/a^*) + \sum_{i=1}^{m} (\beta_i^*/a^*)\, g_{ih} \tag{1.38}$$

where β_0, a, β_i are appropriate Lagrange multipliers with β_0^* associated with $\Sigma\lambda_j = 1$ being free in sign. If the firm is not dynamically efficient, then the inequality sign holds in (1.38), implying less than the optimal growth of output.

Two types of extension of this approach can be easily specified. One is to assume that c_i declines over time due to learning curve effects. Then this can be incorporated in the modified objective function of (1.37) and a new dynamic efficiency frontier computed for the efficient firm. In such cases the transformed model may be a nonlinear program. Note we assume here that output growth is induced by demand growth, hence efficiency improvement is primarily due to innovations on the demand side. When demand falls, the effects are on the reverse side. Thus the general underlying principle is that the successful firms tend to maintain dynamic or growth efficiency over time, while others who lag behind feel the pressure to exit.

A second method to test the growth efficiency of firm h in the LP case is to solve for the optimal values of θ_i and ϕ as follows:

$$\begin{aligned} &\text{Max} \quad \phi - \sum_{i=1}^{m} \theta_i \\ &\text{s.t.} \quad \sum_{j=1}^{N} g_{ij}\lambda_j \leq \theta_i g_{ih}; \quad i = 1, 2, \ldots, m \\ &\qquad \Sigma_j z_j \lambda_j \geq \phi z_h; \quad \Sigma_j \lambda_j = 1, \;\; \lambda_j \geq 0. \end{aligned} \tag{1.39}$$

This allows for a non-radial measure of efficiency, since the efficiency parameter θ_i of each input may vary. In case of radial efficiency, one replaces θ_i by the same θ and the objective function becomes: Max $\phi - \theta$. Sengupta (2003) has recently discussed in some detail the implications of these efficiency measures in technology-intensive industries such as microelectronics and computers and found that the successful firms maintaining their dynamic efficiency tend to increase their market shares over time.

The implications of specific input components such as R&D or knowledge capital may be more clearly specified by a dynamic cost frontier model. Let $\hat{C}_j = \Delta C_j / C_j$ and $\hat{y}_j = \Delta y_j / y_j$. Denote the average growth rates of R&D costs and output or sales of firm j. To test if firm h is dynamically efficient one solve the optimization model

$$\begin{aligned} &\text{Min} \quad \theta \\ &\text{s.t.} \quad \sum_{j=1}^{N} \lambda_j \hat{C}_j \leq \theta \hat{C}_h \\ &\qquad \Sigma_j \lambda_j \hat{y}_j \geq \hat{y}_h; \quad \Sigma_j \lambda_j \hat{y}_j^2 \geq \hat{y}_h^2 \\ &\qquad \Sigma_j \lambda_j = 1, \quad \lambda_j \geq 0. \end{aligned}$$

If firm h is growth efficient, then it must be on the efficiency frontier

$$\hat{C}_h = (a^*/b^*)\hat{y}_h + (\alpha^*/b^*)\hat{y}_h^2 - (b_0^*/b^*) \tag{1.40}$$

where the Lagrangean is

$$L = -\theta + b\left(\theta\hat{C}_h - \Sigma_j\lambda_j\hat{C}_j\right) + a\left(\Sigma_j\hat{y}_j\lambda_j - \hat{y}_h\right) + b_0\left(1 - \Sigma_j\lambda_j^{\partial}\right) + \alpha\left(\Sigma_j\lambda_j\hat{y}_j^2 - \hat{y}_h^2\right).$$

On setting the second order condition as an equality, that is, $\Sigma_j\lambda_j\hat{y}_j^2 = \hat{y}_h^2$, the optimal multiplier value α^* becomes free in sign; hence if α^* is negative, then $\hat{C}_h$ may fall as $\hat{y}_h^2$ rises thus indicating a learning curve and scale effect. In this case the dynamically efficient firm would tend to reap the benefits of declining costs at an increasing rate thus forcing the inefficient firms to exit at a faster rate.

When panel data on inputs and output are available over time $t = 1, 2, \ldots, T$ for an industry, the growth frontier models (1.37) through (1.40) may be easily applied to form two groups of firms: dynamically efficient or not and then apply the replicator dynamics equation (1.33) indiscrete time form:

$$\frac{s_j(t+1)}{s_j(t)} - 1 = \lambda(\bar{c}(t) - \delta c_j(t)) \tag{1.41}$$

where $s_j(t)$ is the relative share of firm j in total industry output in year t, $c_j(t)$ is the average cost per output (or average R&D costs if a specific important input is considered), $\bar{c}_t$ is the mean of $c_j(t)$ over N firms in year t, δ is a dummy variable taking the value 1 if the firm j is dynamically efficient, zero otherwise and λ is the positive speed of adjustment. As Mazzucato (2000) has shown that the change in mean average cost may be related to the variance or variety in efficiency levels across the population:

$$d\bar{c}(t)/dt = -\text{var}(c_j(t)) = -\sigma^2(t)$$

where $\text{var}(\cdot) = \sum_{j=1}^{N} s_j(t)\left(\bar{c}^2(t) - c_j^2(t)\right)$ is the variance. Note that the more general formulation in (1.34) assumes that the fitness or efficiency characteristics may vary so that

$$\bar{c}(t+1) = \alpha_1\bar{c}(t) - \alpha_2\sigma^2(t), \quad \alpha_1, \alpha_2 \geq 0; \; \alpha_1 + \alpha_2 = 1 \tag{1.42}$$

When $\alpha_1 = 1.0$ and $\alpha_2 = \alpha$ one obtains the Fisherian equation (1.34).

From an applied economic standpoint of efficiency analysis the two propositions (1.41) and (1.42) are econometrically verifiable, since δ provides a classification of firms into efficient and not efficient categories. Thus high-tech industries like computers and telecommunications are most likely to have high adjustment rates and higher concentration implying high exit rates when innovations efficiency occurs. Second, one could test from a regression estimate of (1.42) if there is a mean reversion tendency when variance increases over time.

When technological innovations come in diverse forms as in Schumpeter's theory, the variance term $\sigma^2(t)$ may change over time due to heterogeneity. In this case one may have to apply an ARCH (autoregressive conditional heteroscedasticity) type model, for example,

$$\sigma^2(t) = a_0 + a_1\sigma^2(t-1) + \varepsilon(t-1)$$

where $\varepsilon(t)$ is a zero mean error component. On using conditional estimates of $\sigma^2(t)$ and $\varepsilon(t)$ the variance model may be specified as

$$\sigma^2(t) = \beta_0 + \beta_1\hat{\sigma}^2(t-1) + \beta_2\hat{\varepsilon}^2(t-1) + \eta(t-1).$$

This model provides a simple framework for analyzing the persistence of heterogeneity in the cost output data due to heteroscedasticity. A second reason is that this specification allows a direct test of the *mean reversion hypothesis* of the efficiency distribution of firms in terms of costs. This hypothesis tests that the higher deciles of the efficiency distribution generate smaller increases in efficiency than the lower deciles.

This idea has been applied by Mazzucato (2000) in a series of simulation studies based on a differential equation model

$$\frac{c_j(t+1)}{c_j(t)} - 1 = -\lambda(1 - s_j(t)), \quad \lambda > 0.$$

This assumes that the speed of cost reduction of an efficient firm j falls as its market share increases. The adjustment parameter λ determines the speed with which intra-industry costs converge to a minimum cost. It is an industry-specific parameter which can be interpreted as the strength of spillover of knowledge and diffusion, that is, the degree of technological opportunity available. By making λ depend on both firm-specific innovations (h_j) and industry-specific knowledge capital (H), that is, $\lambda = \lambda(h_j, H)$ one could profile a wide variety of cost and hence efficiency trajectories of firms under dynamic competition. The detailed simulation studies by Mazzucato show four types of distinct trajectories:

1. For value of $\lambda < 0.002$ a monopolistic type of market emerges, because costs change very slowly and the selection process completely dominates the evolutionary process thus ensuring only the initially most fit firms to survive.
2. For values $0.002 < \lambda < 0.007$ unit costs fall slowly but rapidly enough to allow partial coexistence of those near-efficient firms whose costs converge before the selection mechanism has time to force all laggard firms out of the industry.
3. For values $0.007 < \lambda < 0.030$ a switching pattern among market shares emerges. This closely agrees with the empirically observed trend where market share turbulence (i.e., increasing $\sigma^2(t)$) tends to be higher during periods in which small firms have relative comparative advantages in the early phase of the innovations process.
4. For a given value of λ in the range $0 < \lambda < .03$ but with different variance levels (i.e., 0.004, 0.200 and 0.400) for the initial distribution of costs, all with mean 0.6 it is found that with any given value of λ, the higher the initial variance of costs, the longer it takes for market shares to reach an equilibrium value.

These simulation results provide interesting insights into the feedback relation between firm size and innovations. First of all, the stochastic forces associated with the Schumpeterian process of 'creative destruction' and also the R&D process viewed as research creativity and learning by doing have a dynamic role in the equation (1.41) of growth. Secondly, the cost reduction process, which is called the dynamic increasing returns by Mazzucato occurs at diverse rates for different rates, thus increasing the comparative advantages of the successful firm and decreasing the same for the laggards. This may be seen more clearly in equation (1.40), which implies that the average cost of expansion defined as $\hat{c} = \hat{C}/\hat{y}$ may be minimized by choosing an optimal level of expansion

$$\hat{y}^* = (-\alpha^*/b^*)^{1/2} \tag{1.43}$$

where α^* is negative due to declining unit costs through dynamic increasing returns. The leaders among growth-efficient firms would follow trajectories along (1.43). This may involve a continual process of innovative activities both creative and diversifying.

1.5.2 Innovations and market dominance

In Schumpeterian framework innovations tend to provide several channels of potential market power which may deter future entry. For this purpose consider first a Cournot type model, where each firm j seeks to maximize profits

$$\pi_j = p(y_j + Y_{-j})y_j - C(y_j) \tag{1.44}$$

where y_j is the output of firm j, Y_{-j} is the total output of all its rivals, C is total cost of production with marginal cost $\hat{c}_j$, and p is the industry price depending on total output $Y = y_j + Y_{-j}$ which equals total demand. Maximizing profits π_j by each firm yields the optimal mark-up μ

$$\mu = (p - \hat{c}_j)/p = (1 + a_j)s_j/e \tag{1.45}$$

where $s_j = y_j/Y$ is the market share, $a_j = (\partial Y_{-j}/\partial y_j)$ and the price elasticity demand is $e = (p/\text{Y})(\partial Y/\partial p)$. To arrive at the average mark-up of price over marginal cost in an industry we weight the above equation by the market share and derive

$$\sum_{j=1}^{N} s_j^2(1 + a_j)/e = (p - \bar{c})/p$$

where $\bar{c} = \sum_{j=1}^{N} s_j\hat{c}_j$, $\hat{c}_j = \partial C(y_j)/\partial y_j$. If the reaction coefficients a_j are identical $a_j = a$, then this yields the result

$$\bar{\mu} = (p - \bar{c})/p = (1 + a)H$$

where $H = \sum s_j^2$ is the Herfindahl index which is an index of concentration. Since the left hand side can be viewed as a profit sales ratio, it is clear that increasing H yields increased profit–sales ratio; also one can view a_j as a measure of effective collusion, so that increased degree of collusion increases profit margins. Note that $a_j = \partial Y_{-j}/\partial y_j$ is the perceived response of rival firms to a change in output of firm j and hence this may depend on the concentration index H also. The Schumpeterian theory of innovations assumes a stream of innovations whereby a successful innovator produces the newly invented good and protected by the patent right drives out the previous incumbent by undercutting his price and enjoys monopoly rents until driven out by the next innovation. Thus the success in driving out the previous incumbent depends on his ability to reduce costs by new innovations and thereby creating a local monopoly. Since the rate of change in market share can be viewed as proportional to the cost differential, that is,

$$\dot{s}_j/s_j = \lambda(\bar{c} - c_j); \quad c_j = \text{unit cost} \tag{1.46}$$

as in equation (1.41) before, one could also write (1.45) as

$$\dot{\mu} = [\lambda(1 + a_j)s_j(\bar{c} - c_j)/e]$$

assuming λ, a_j and e to be fixed. This shows that profit margin increases over time, if unit costs c_j can be lowered by innovations. Here λ determines the speed of selection: the speed at which firm shares react to differences between firm efficiency characteristics. The speed of selection could be made a firm- or an industry-specific parameter, which may also evolve over time. As an industry-specific parameter a high λ would describe an industry with a strong competitive adjustment mechanism, which may evolve endogenously with the changing Herfindahl index. A low λ likewise would represent monopolistically competitive adjustment. In the latter case excess profits or returns are competed away more slowly due to higher concentration ratio. Kessides (1990) has shown empirically for US manufacturing industries that the following tendencies persist:

$$\frac{\partial \lambda}{\partial H} > 0, \quad \frac{\partial \lambda}{\partial g} > 0, \quad \frac{\partial \lambda}{\partial \text{MES}} > 0 \quad \text{and} \quad \frac{\partial \lambda}{\partial K} > 0 \tag{1.47}$$

where g is the growth rate of industry demand, MES is a measure of minimum efficient scale of output and K represents total capital required for an MES firm. The first states that excess returns are competed away more slowly in industries that are highly concentrated. The second says that excess profits erode more rapidly in industries experiencing slow growth, where the latter enhances the destabilizing effects of entry. The third hypothesis says that excess profits tend to erode slowly in industries with strong scale economies and finally, the last hypothesis says that excess profit erodes slowly in industries with high absolute capital requirements.

The role of H, G, MES and K in dynamic competition are considered central to the model of hypercompetition developed D'Aveni (1994). Sengupta (2002) has discussed this in some detail in Schumpeterian framework, where he stressed three types of efficiency: technological efficiency, access efficiency and resource (stronghold) efficiency. Resource efficiency is emphasized by the contribution of K in (1.47), technological efficiency by MES and the access efficiency by H and g. By building barriers around a stronghold, the firms reap monopoly profits in a protected market that can be used to fund aggressive price strategies and R&D investments. Access to distribution channels and low-cost supply sources in the supply chain and also product differentiation provide along with dynamic economies of scale the major barriers to entry that the successful innovator firms use to create and sustain a stronghold.

Since increased market share may be viewed as entry, it is clear from (1.46) that potential entry could be reduced by increasing efficiency so as to reduce unit costs c_j below the industry cost level $\bar{c}$. Streams of innovation flows in Schumpeterian framework are intended to play this role. In a general setting one may formalize the entry dynamics by a two-equation model for an industry

$$\dot{E} = k(\pi - \pi^*)$$
$$\pi^* = h(B) \tag{1.48}$$

where $\dot{E}$ is entry, π is expected profitability with its long run value π^* and B is the set of barriers to entry from different channels such as advertisement, dynamic economies of scale and the three forms of Schumpeterian efficiency, that is, technological, access and stronghold efficiency. The role of the incumbent and innovating firms in this framework (1.47) has to be worked out in terms of cost price implications and market shares. Sutton (1998) has developed in this framework game-theoretic models emphasizing the role of endogenous technical progress.

Consider for example a single R&D trajectory. There are N_0 firms indexed by j. At any time t firm j has a level of technical capability denoted by $u_j(t)$ with its maximum value $\bar{u}(t)$ interpreted as the quality of firm j's product. We define the cost of R&D investment at time t which will lead to a level of technical capability $k\bar{u}(t)$ by time $t + T$ as $F(k, T)$ where it is assumed

$$F(k, T) = k^{\beta} g(T), \quad \beta > 0 \tag{1.49}$$

with $f(T)$ decreasing in T.

This implies that speeding up the R&D project is costly. Associated with each N-tuple $(u_j(t))$ there is a flow of profit per unit time for each firm. Thus profits earned by a firm j in interval $(t, t + \mathrm{d}t)$ for a sequence of investments $(u_j(\mathrm{d}t))$ is given by

$$S\pi(u_j(t)|u_{-j}(t))\ \mathrm{d}t \tag{1.50}$$

where S is a measure of market size and $u_{-j}(t)$ denotes all other rival firms. We assume that a rise in any rival firm's technical capability reduces or leaves unchanged the profit stream of firm j.

The model specified by (1.49) and (1.50) then maps the configuration $(u_j(t))$ into an action or strategy space with a corresponding payoff for each firm, which we may define as the net present value of the profit stream less the R&D costs incurred, discounted to time $t = 0$ at some positive discount rate.

The implications of variables such as g, MES, K and H as outlined in (1.47) may then be easily worked out in a game-theoretic context. Cabral and Riordan (1994) have considered a closely related game-theoretic model which posed the question: Once ahead, what does the leading firm have to do to stay ahead? Here it is assumed that any firm's unit cost c(s) is a decreasing function of cumulative past sales s. Firms maximize expected discounted costs and the solution concept used is a sub-game Markov-perfect equilibrium, where each firm's strategy depends only on the state of the game. Clearly the market dominance of the leading firm would depend on two dynamic effects: a cost effect and a strategy effect. Thus a firm at the very bottom of its learning curve maintains a strategic advantage as long as its rival has a higher cost. The strategic or prize effect refers to the potential prize from winning the lagging firm. Thus if the lagging firm has a sufficiently larger prize, then the prize effect could dominate the cost effect. In equilibrium the price difference of two firms is proportional to the cost difference, that is, $p_2 - p_1 = k(c_2 - c_1)$ where c_1, c_2 are the unit costs of two firms.

Two points are to be noted in the formulation. One is that the factor of proportionality k in the price differential would be higher, the higher the market dominance of the leading firm. Second, the learning curve effects assumed here are finite and each firm reaches the bottom of its learning curve upon making m sales $(m > 1)$. This second assumption is at variance with the Schumpeterian framework of streams of innovations.

Note

1. The objective function of (1.11) may be written as: Max $C_h^* = \gamma_0 + \gamma_1 y_j + \gamma_2 y_j^2$ so that the dual problem may be written as a minimization problem.

References

Athanassopoulos, A. and Thanassoulis, E. (1995): Market efficiency from profitability and its implications for planning. *Journal of the Operational Research Society* 46, 20–34.

Baumol, W. (1967): *Business Behavior, Value and Growth*. Harcourt, Brace and World, New York.

Cabral, L. and Riordan, M. (1994): The learning curve, market dominance and predatory pricing. *Econometrica* 62, 1115–40.

Charnes, A., Cooper, W., Lewin, W. and Seiford, L. (1994): *Data Envelopment Analysis*. Kluwer Academic Press, Boston.

D'Aveni, R.A. (1994): *Hypercompetition: Managing the Dynamics of Strategic Maneuvering*. Free Press, New York.

Dosi, G. (1984): *Technical Change and Industrial Transformation*. Macmillan Press, London.

Downie, J. (1958): *The Competitive Process*. Duckworth, London.

Dreze, J. and Sheshinski, E. (1984): On industry equilibrium under uncertainty. *Journal of Economics Theory* 33, 88–97.

Farrell, M.J. (1957): The measurement of productive efficiency. *Journal of Royal Statistical Society*, Series A 120, 253–90.

Fisher, R.A. (1930): *The Genetical Theory of Natural Selection*. Clarendon Press, Oxford.

Gabszewicz, J. and Michel, P. (1991): Capacity adjustments in a competitive industry. In: Gabszewicz, J. (ed.), *Equilibrium Theory and Applications*. Cambridge University Press, Cambridge.

Gaskins, D.W. (1971): Dynamic limit pricing: Optimal pricing under threat of entry. *Journal of Economic Theory* 3, 306–22.

Hall, R.E. (1990): Invariance properties of solow's productivity residual. In: Diamond, P. (ed.), *Growth, Productivity and Unemployment*. MIT Press, Cambridge, MA.

Johansen, L. (1972): *Production Functions*. North Holland, Amsterdam.

Jovanovic, B. (1982): Selection and the evolution of industry. *Econometrica* 50, 649–70.

Kessides, I.N. (1990): The persistence of profits in US manufacturing. In: Mueller, D.C. (ed.), *The Dynamics of Company Profits*. Cambridge University Press, Cambridge.

Lansbury, M. and Mayes, D. (1996): Entry, exit, ownership and the growth of productivity. In: Mayes, D. (ed.), *Sources of Productivity Growth*. Cambridge University Press, Cambridge.

Lucas, R.E. (1993): Making a miracle. *Econometrica* 61, 251–72.

Mansfield, E. (1962): Entry, Gibrat's law, innovations and the growth of firms. *American Economic Review* 52, 1023–51.

Mazzucato, M. (2000): *Firm Size, Innovation and Market*, Edward Elgar, Cheltenham.

Metcalfe, J.S. (1994): Competition, evolution and the capital market. *Metroeconomica* 4, 127–54.

Nelson, R. and Winter, S. (1982): *An Evolutionary Theory of Economic Change*. Harvard University Press, Cambridge, MA.

Novshek, W. (1980): Cournot equilibrium with free entry. *Review of Economic Studies* 47, 473–86.
Schumpeter, J. (1942): *Capitalism, Socialism and Democracy*. Harper and Row, New York.
Sengupta, J.K. (1990): Transformations in Stochastic DEA models. *Journal of Econometrics* 46, 109–23.
Sengupta, J.K. (2000): *Dynamic and Stochastic Efficiency Analysis*. World Scientific, London.
Sengupta, J.K. (2002): Model of hypercompetition. *International Journal of Systems Science* 33, 669–75.
Sengupta, J.K. (2003): *New Efficiency Theory*. Springer-Verlag, Berlin.
Sutton, J. (1998): *Technology and Market Structure*. MIT Press, Cambridge, MA.
Timmer, C.P. (1971): Using a probabilistic frontier production to measure technical efficiency. *Journal of Political Economy* 79, 776–94.

2
Selection and Evolution of Industry

2.1 Introduction

The evolution of industry depends on the selection mechanism, that is, the process of entry and exit of firms and the various factors influencing the entry and exit decisions. Three major factors are most important in the selection process. One is the evolutionary perspective which emphasizes the firm's ability and competence to alter their market position and hence through strong increasing returns to scale, to alter the market structure significantly. Following the Schumpeterian theory of technological innovations the main source of such increasing returns is the cumulative aspect of innovations, where "size begets size" causes industrial dynamics to be characterized by nonlinear and path dependent processes, where random events like a new technical process may have lasting and irreversible effects on the dynamic evolution of the selection process. Secondly, firms differ significantly in their commitment and ability to innovate. Thus innovations in products and processes are largely endogenous to the firm through R&D investment and learning-by-doing. Thus vigorous innovation has been found to generate more competitive market structures, while innovations requiring large investment generally involve more concentration through large size firms. Also due to the cumulative nature of technological innovations firms that discover new technologies are able to maintain their lead even after the particular technology is obsolete. This is specially true if network externalities cause consumers and other firms to "lock into" a particular technology, so that even when a new technology appears, it may not be adopted for some time. Finally, the evolutionary forces of selection which allow only some firms to survive and grow are subject to initial chance events and stochastic mechanisms, which play an active dynamic role. Thus Jovanovic (1982) and Mazzucato (2000) have considered cost-efficiency depending on a stochastic parameter. In economic terms uncertainty measured by the stochastic parameter is generally very high in the early stages of the industry life cycle particularly when the product design has not yet been standardized. In this phase, the flexibility of small new firms allows them to be the leaders in cost reduction and radical innovations thus causing high rates of entry. During the mature stage of the industry life cycle however, economies of scale and learning-by-doing favoring large sized firms are very strong and the innovation becomes increasingly path dependent, thus yielding a more stable oligopolistic structure. Klepper (1996) and others have shown that innovation could lead to increased concentration if successful innovators rise to market dominance, for example, if an innovation caused the minimum efficient scale (MES) of production to grow more rapidly than demand, then the concentration measured by the share of the largest firms would increase. Empirically this has been found to hold in chemical industries, electric power generation, cement and others.

The entry process may be viewed in two broad ways. One is in terms of the entrant's output or price behavior vis-à-vis the incumbent. For example, if y_1 and y_2 are the two output shares of industry output, then changes in y_1 may depend on the cost price strategy of the incumbents or the innovation efficiency (I_1) of the entrants which reduces their average cost (c_1) relative to the industry average $\bar{c}$, that is,

$$dy_1/dt = f(\bar{c} - c_1,\ I_1) \tag{2.1}$$

where $c_1 \leq \bar{c}$ is assumed. The exit process may be similarly viewed, when $c_2 > \bar{c}$ or the incumbents fail to maintain dynamic innovation efficiency. Secondly, the price (p) quantity (q) adjustment process may reflect the entry–exit dynamics:

$$\dot{q} = a[p - c(q)]; \quad \dot{p} = b[D(p) - q]. \tag{2.2}$$

Here dot denotes time derivative, $c(q)$ is average cost, $D(p)$ is demand and a, b are positive constraints reflecting speed of adjustments. This type of entry prices assumes that excess profits invite more entry and exits follow when excess profits become negative.

2.2 Entry and exit processes

One of the major determinants of the entry process is the existence of potential and actual barriers. The pioneering work by Bain (1956) argued that firms can earn profits above the competitive level in the long and medium term, if they are protected by "*entry barriers*". He identified four main types of entry barriers, for example, economies of scale, product differentiation advantages, absolute cost advantages and large capital stock or fixed input requirement. To this list one may add two more: innovations efficiency of the incumbents in the timing and know-how arena and the access efficiency, where the companies keep potential entrants out by competing aggressively and constantly moving forward, for example, Coke and Pepsi. These latter two barriers have been stressed in the current theory of hypercompetitive markets, closely following in the Schumpeterian tradition.

Bain considered economies of scale as an important barrier to entry. If the incumbent firm operates a MES plant and the new entrant can come into the industry only at less than MES capacity, then the entrance is not feasible. The established firms may then engage themselves in limit pricing. Here the MES plan refers to the output level at which average cost is minimized. Stigler (1968), Rees (1973) and others used the MES concept to explain the pattern of entry and exit by the principle of survival of the fittest. If the proportion of industry output produced by a particular size class of firms decreases over time, then it must be an inefficient class. A size group that survives for a long time or does not reduce its market share is efficient. However if the cost functions are different or there is product differentiation, the survivorship principle can only reveal the range of efficient plant sizes.

Product differentiation advantages of the incumbents could retard entry either due to brand loyalty by consumers or the patent control over the quality of products. An absolute cost advantage barrier may arise due to either the control of supply of vital raw materials or know-how or software by the established firms, or the control of the efficient productive technique by the established firms.

The entry and exit dynamics may be modeled in four different ways based on (a) profit rates (b) price and quantity adjustments (c) Cournot–Nash type adjustments and (d) game-theoretic constructions. We now discuss these alternative ways of selection dynamics.

In respect of the profit based adjustment, we may refer to the econometric study by Orr (1974) who estimated a model of the form

$$\Delta n = \mathrm{a}(r - r^*) \tag{2.3}$$

where Δn is the gross increase in the number of sellers over the period, a is a positive constant measuring the speed of adjustment, r is the average profit rate and r^* is the profit rate at which entry would stop, that is, entry preventing profit rate. Since r^* is not observable, he substituted a linear function of variables representing entry barriers and other entry conditions. His study based on Canadian manufacturing data for the period 1964–67 confirmed the Bain conclusion that high barrier industries are significantly more profitable than other industries.

Following this line of analysis Salinger (1984) argued that Bain's hypothesis implies that r depends on the various structural determinants of entry (E_i) and past sales growth, for example,

$$r = \alpha_0 + \alpha_1 \text{CCN} \cdot E_0 + \alpha_2 G \tag{2.4}$$

$$E_0 = \beta_0 + \beta_1 E_1 + \beta_2 E_2 + \cdots + \beta_n E_n$$

Here CCN is a measure of industry concentration (e.g., share of top 10 firms in industry output), E_0 is the composite entry barrier index and G is a measure of past sales growth. This equation implies that without barriers to entry there is no reason for market power to rise just because the industry is more concentrated. One has to note however some basic weakness of these empirical studies. For example, high entry barriers may imply modest profits, if the rents of scarce factors have been capitalized. Also the situation $r > r^*$ may not be very persistent if there exists stochastic market uncertainty. Uncertainty tends to differ between industries and between different periods in the life of an industry. The analysis of the life cycle is particularly important here. During the early stage of an industry's evolution, when entry rates are high and the market is uncertain, the flexibility of small new firms allows them to be the main source of cost reduction and innovation thus causing high rates of entry. A "shakeout" tends to occur after the product price declines. Then more stable market demand sets in, thus allowing larger firms to reap economies of scale and utilize process innovation.

The price and quantity adjustment processes under competitive framework may be of different forms. One useful form studied by Dreze and Sheshinski (1984) assumes that each firm in an industry produces a single product but chooses one of the K possible cost structures. If q_i is the output of firm i and $F_i(q_i)$ denotes its total cost assumed to be convex, then its average cost is assumed to follow a standard U-shaped curve with a minimal point at q_i^*. Let n_i be the number of firms of type i, $i = 1, 2, \ldots, K$, then the total costs C for the industry are given by

$$C = \sum_{i=1}^{K} n_i F_i(q_i). \tag{2.5a}$$

Given the total demand D the industry selection process minimizes C in (2.5a) subject to the condition that production satisfies demand,

$$\sum_{i=1}^{K} n_i q_i \geq D. \tag{2.5b}$$

Let $\hat{q}_i = \hat{q}(p(n, D))$ be the optimal solution for a given positive demand, when p is the optimal Lagrange multiplier for (2.5b) and n is the vector $(n_1, n_2, \ldots n_K)$ of number of firms. The optimal industry cost is then

$$L(n) = C(n, D) = \sum_{i=1}^{K} n_i F_i(\hat{q}_i(n, D)).$$

It is easy to show that $L(n)$ is a convex function of n; hence there exists a vector n^* at which $L(n)$ attains a minimum. Clearly n^* satisfies the conditions

$$\psi_i(n^*) \leq 0 \text{ and } n_i^* \Psi_i(n^*) = 0, \text{ all } i. \tag{2.5c}$$

They define competitive equilibrium by the number of firms $\hat{n}$, the prices $\hat{p}$ and quantities $\hat{q}_i$ which equalize total demand and supply and also maintain the conditions (2.5c) for minimum industry costs.

The dynamic process P for entry and exit of firms is then modeled as

$$\dot{n}_i = \begin{cases} g_i(\Psi_i), & n_i > 0 \\ \max(0, g_i(\Psi_i)), & n_i = 0 \end{cases} \tag{2.6}$$

and for all t, the price $p(t)$ equalizes demand and supply. Here $\dot{n}_i = dn_i(t)/dt$ and the functions $g_i(\psi_i)$ are assumed to be continuous, strictly increasing and bounded with $g_i(0) = 0$.

Dreze and Sheshinski have proved an important theorem.

Theorem (Dreze and Sheshinski). Under the assumptions of a standard U-shaped average cost curve for each firm and a continuous, non-increasing demand function D, the entry–exit process P in (2.6) is quasi-stable, that is, if for any initial position (n_0, t_0) the solution $n(t; n_0, t_0)$ is bounded and every limit point of $n(t; n_0, t_0,)$ is an equilibrium point as $t \to \infty$.

A number of comments may be added here. First of all, the competitive equilibrium defined by the conditions (2.5c) for $\hat{n} = n^*$ and the corresponding $p(n^*, D)$ and $q_i(n^*, D)$ satisfies for all $i = 1, 2, \ldots, n_K$ the maximizing profit conditions

$$\psi_i(n^*) \leq 0 \text{ and } n_i \psi_i(n^*) = 0, \quad \text{all } i$$

where $\psi_i(n) = q_i(MC_i - AC_i)$ is the profit of a firm of type i. Thus $q_i^* = \hat{q}_i$ is an optimal supply for firm i given the conditional prices $p^* = p(n^*, D)$, where optimal supply is still defined by maximization of profits. Thus price quantity adjustment is implicit in the entry–exit process P. In other words, we assume in the process P defined by (2.6) that new plants of type i are built (scrapped) whenever their profits are positive (negative). This formulation is an idealized representation of the investment process.

Second, the number n_i denotes the number of firms which must be integers. One has to interpret these variables as continuous proxies to be discrete variables N_i defined by $N_i = [n_i]$, where $[k]$ for any real number k denotes the greatest integer that does not exceed k. Third, if we assume that the equilibrium number of firms n^* defined by the stationary point of the entry–exit process in (2.6) is unique, then the distance between the solution path $n(t; n_0)$ and n^* may be used as a Lyapunov function

$$V = \left(\frac{1}{2}\right)[n(t, n_0) - n^*]^2.$$

It then follows that

$$\dot{V}_t = g_0(n(t, n_0) - n^*)\psi_i(n^*)$$

is negative for $n(t, n_0) \neq n^*$, where g_0 is the linearized slope of the function $g(\psi_i)$. Therefore it follows that $\lim_{t\to\infty} n(t, n_0) = n^*$.

Thus suitable liberalizations of the entry–exit process $g_i(\psi_i)$ around the stationary or equilibrium point n^* provide a domain of local stability. But for multiple equilibria the P process maybe only quasi-stable.

A more direct form of Walrasian price–quantity adjustment involving an entry–exit process may be introduced by a system of differential equations for quantity q and price p, where each firm has an average cost function $c(q)$ and the total demand function if $D(p)$.

$$\dot{q} = a[p - c(q)], \quad a > 0 \tag{2.7}$$
$$\dot{p} = b[D(p) - q], \quad b > 0.$$

The equilibrium values p^* and q^* are defined by $\dot{q}$ and $\dot{p} = 0$. On linearizing the nonlinear system (2.7) around (q^*, p^*) and evaluating the characteristic equation we get

$$\lambda^2 + \lambda(ac' - bD') + ab(1 - c'D') = 0 \tag{2.8}$$

where $c' = \mathrm{d}c/\mathrm{d}q$ and $D' = \mathrm{d}D/\mathrm{d}p$ are the slopes of the average costs and demand functions. If demand has a negative slope $D' < 0$ and the average cost function is rising $c' > 0$, then the two roots of (2.8) have negative real parts and hence the Walrasian adjustment process (2.7) is stable around the equilibrium point (q^*, p^*), if it is assumed to be unique.

Note however that if the conditions $D' < 0$ and $c' > 0$ are not fulfilled in some domain, this type of local stability in the sense of convergence of $q(t)$ and $p(t)$ to q^* and p^* respectively may not hold.

2.2.1 Cournot–Nash adjustments

We discuss in this section Cournot–Nash equilibrium in a simplified framework, where each firm produces a single homogeneous good q_i with an industry aggregate output $Q = \sum_{i=1}^{n} q_i$ for n firms. The market clearing price is $p = f(Q)$ and the cost function $C(q_i)$ is assumed to be identical. The more general case of cost structures may be easily generalized. The demand function is assumed to be continuous with a negative slope and the cost function is assumed to be strictly convex and continuous. The profit function of firm i is then defined by

$$\pi_i = \pi_i(q_i, Q_{-i}) = q_i f(q_i + Q_{-i}) - C(q_i) \tag{2.9}$$

where $Q_{-i} = Q - q_i$. The Cournot–Nash (CN) equilibrium is then defined by

$$f(Q(n)) + q_i f'(Q(n)) - C'(q_i(n)) = 0, \quad i = 1, 2, \ldots, n \tag{2.10a}$$

where prime denotes derivatives. Suzumura (1995) has considered a Cournot–Nash type adjustment dynamics around the equilibrium solution of (2.10a). He assumed that a symmetric and positive CN equilibrium uniquely exists for each n, so that at $q_i(n) = q(n)$ for all $i = 1, 2, \ldots, n$. The entry dynamics is formulated by the differential equation

$$\dot{q}_i = a\delta\pi_i/\delta q_i, \quad a > 0, \quad i = 1, 2, \ldots, n \tag{2.10b}$$

where a is a positive coefficient denoting speed of adjustment and $\dot{q}_i$ is the time derivative of output. Let $\pi(n)$ be the profits of each incumbent firm at the CN equilibrium $q(n)$. Then if $\pi(n) > 0$ holds, potential competitors will be induced to enter into this profitable industry in the long run. In the opposite case $\pi(n) < 0$ and the incumbent firms may be forced to exit in the long run from this unprofitable industry. The dynamic entry–exit mechanism may then be formulated by the differential equation

$$\dot{n} = b\pi(n), \quad b > 0. \tag{2.10c}$$

Let n^* be the equilibrium number of firms defined by $\pi(n^*) = 0$. Let $n(t; n_0)$ be the solution path of equation (2.10c) starting from the initial value $n_0 > 0$. Once again we may use the Lyapunov distance function V between the solution path $n(t; n_0)$ and n^*

$$V = \frac{1}{2}[n(t, n_0) - n^*]^2 \tag{2.10d}$$

which on differentiation becomes

$$\dot{V} = \mathrm{d}V/\mathrm{d}t = b(n(t, n_0) - n^*)\pi(n(t, n_0))$$

which is negative as long as $n(t, n_0) \neq n^*$. Thus there exists a stable convergence $\lim_{t\to\infty} n(t, n_0) = n^*$.

Note that if the entry process (2.10c) is written in a nonlinear form

$$\dot{n} = g(\pi(n))$$

as in (2.6) before, then the stability results may not hold for all manifolds. In particular there may be multiple equilibrium points.

Suzumura (1995) has developed two types of entry–exit processes. One is given by (2.10c). The other is based on the concept of a socially first-best output profile $q^F(n)$ defined by minimizing the market surplus function defined by the sum of consumers' and producers' surplus W.

$$W = \int_0^Q f(z)\,\mathrm{d}z - \sum_{i=1}^{n} C(q_i). \tag{2.11a}$$

Assuming interior optimum the output vector $q^F(n) = ((q_i^F(n), \ldots, q_n^F(n))$ can be characterized by

$$(\partial/\partial q_i)W = f(Q^F(n)) - C'(q_i^F(n)) = 0.$$

Then he shows that $q^F(n) > q(n)$ holds for any n, where $q(n) = (q_i(n))$ as in (2.10a). Let n_f be the first-best number of firms maximizing W in (2.11a) and $n = n_s$ be the second-best number of firms satisfying (2.10a). Then n_s exceeds the socially first-best number of firms $n_{f,}$, whenever $n_f > 1$. Furthermore,

$$f(Q^F(n_f)) = C(q^F(n_f))/q^F(n_f) \tag{2.11b}$$

that is, price equals average cost. As an example consider a linear demand and quadratic cost function model

$$p = A - Q, \quad C(q) = K + aq + (b/2)q^2;$$

then it can be shown that

$$q^F(n) - q(n) = (A - a)/[(n + b)(n + b + 1)]$$

which shows that $\lim_{n\to\infty} (a^F(n) - q(n)) = 0$, that is, in the limit the Cournot equilibrium output $a(n) = q^s(n)$ converges to the first-best (perfectly competitive) output $q^F(n)$ as n tends to infinity. Note that $\pi(n) = \pi^s(n) > 0$ if and only if $n < n^*$ and vice versa. Therefore in the presence of positive fixed cost K, which may generate economies of scale, there exist no private incentives for firm entry beyond n^* given by (2.10d).

One important welfare implication of the second-best excess entry theorem (i.e., $n_s > n_f$) at the margin is that it is welfare-improving if we marginally decrease the number of firms from its free-entry equilibrium level.

The above two results on the first-best and second-best excess entry theorems may be generalized in the case where each firm chooses one of several cost functions as in the Dreze–Sheshinski model. But the existence of unequal fixed costs for different firms may alter the long-run pattern of profits. For example, if c_i is the average cost of firm i and it declines

$$\dot{c}_i/c_i = -\alpha(s_i), \quad \alpha > 0 \text{ for each } s_i > 0$$

when the output size s_i of firm i increases, then this may generate a reduction in price causing inefficient firms to exit at a faster rate. For dynamically inefficient firms the cost reductions are either nonexistent or negative.

2.2.2 Game-theoretic models

Game-theoretic formulations provide a more direct analysis of the interdependence and rivalry among firms, which need not be symmetric in the strategy space. Gaskins (1971) and others have developed the model of a dominant firm under conditions of potential entry by the new rivals. The rate of entry into the industry is assumed to depend on current price only and entry is defined by an increase in output from the competitors, comprising the existing rivals and the new entrants. If the barriers to entry are high, the dominant firm can produce the short-run profit maximizing level of output with little fear of losing its market share. However, if entry is relatively easy, the firm can increase its output to the point where the price induces no entry. This price is called the limit price, that is, entry-preventing price.

The above view of entry however is too deterministic. In most realistic situations the dominant firm has imperfect knowledge of the market and the uncertain state of potential entry. Also in the dynamic situation market tends to grow or decline. We consider here a dynamic limit price model under conditions of stochastic entry and a positive rate of growth of market, which represent most of the high-tech industries like computers and telecommunications today. Thus the objective of the dominant firm producing output $q(t)$ is to maximize the present value of risk-adjusted expected profits under conditions of stochastic entry:

$$\begin{aligned}
\text{Max}\quad & J = \int_0^\infty \exp(-rt)[E\pi(t) - m \text{ var } \pi(t)]\, dt \\
\text{s.t.}\quad & \dot{x}(t) = k[Ep(t) - p_0] \qquad\qquad (2.12a)\\
& p(t) = a \exp(nt) - b_1(q(t) + x(t)) + u \\
& \pi(t) = (p(t) - c)q(t).
\end{aligned}$$

Here $q(t)$ is the output of the dominant firm, $x(t)$ the output of the rivals on the fringe, n is the growth rate of market demand, $\pi(t)$ is the profit level with expectation $E(\cdot)$ and variance $\text{var}(\cdot)$ and m is a positive risk aversion parameter. On applying Pontryagin's maximum principle Sengupta and Fachon (1997) have derived the optimal paths of two output variables. In terms of the transformed variables $X(t) = x(t)\exp(-nt)$, $Q(t) = q(t)\exp(-nt)$ these two paths may be written as

$$\begin{aligned}
\dot{X}^*(t) &= -(kb_1 + n)X^*(t) - kb_1Q^*(t) + k(a - p_0 \exp(-nt)) \\
\dot{Q}^*(t) &= (2b_2)^{-1}(2kb_1^2 + rb_1)X^*(t) + [b_1k + r - n]Q^*(t) - A(t) \qquad (2.12b)
\end{aligned}$$

where

$$A(t) = (2b_2)^{-1}[a(2kb_1 - n + r) - (b_1kp_0 - b_1kc - rc)\exp(-nt)]$$

where dot is the time derivative and asterisk denotes an optimal trajectory. These two equations (2.12b) may be interpreted as the optimal reaction curves of the dominant incumbent firm and the potential or existing rivals. At the steady state $\dot{X}^* = 0 = \dot{Q}^*$, the long-run market share of the dominant firm is given by

$$S = n(kb_1 - n + r)[rkb_1 - n^2 + nr + 2k(kb_1 - n + r)mv]^{-1}$$

where $\text{var } \pi(t) = vq^2(t)$.

Clearly it follows that whenever $r < n$, a decrease in long-run market share will result from an increase in uncertainty indexed by the parameter v, an increase in the adjustment coefficient (k) or a decrease in the market growth rate (n).

The characteristic equation of the differential system (2.12b) can be derived as

$$\lambda^2 + (2n - r)\lambda + (n^2 - nr - b_1 kr/2) = 0$$

which implies that the optimal output paths would converge without oscillations to the steady state whenever it holds that

$$|2n - r| > |(r^2 + 2b_1 kr)^{1/2}|.$$

Otherwise there may exist domains of unstable manifolds, as we have seen before. Some simulation results by Sengupta and Fanchon show that a firm facing a very low adjustment coefficient k will rapidly lose market share and the market price will increase nearly at the same rate as the growth of demand. In such a case the position of price leadership of the firm will be quickly challenged by the rivals. In this model the equilibrium price lies somewhere between the short-run monopoly price and the competitive price p_0, the exact positioning depending on the barriers to entry, the degree of risk aversion of the dominant firm and the response coefficient k of the entry equation.

The role of fixed cost or capacity constraint is not explicitly introduced in the dynamic limit pricing model above, although it has significant influence in reducing average costs and hence the limit price. An interesting model of sequential entry has been formulated by Spulber (1981) to discuss the role of capacity in Cournot–Nash type models. Let firm 1 be the established firm and firm 2 the potential entrant. Let q_1^1, q_2^1 denote the outputs of two firms in period 1 and q^2 be the entrant's output in the second period. The variable costs in each period are $c^1(\cdot)$, $c^2(\cdot)$, assumed to be convex, differentiable and increasing. Each firm purchases capacity k^1, k^2 at a fixed price h. For a monopolist facing no threat of entry the maximization of short-run profits yields the optimal capacity k_m, such that the discounted sum of marginal returns equals the cost h of capacity, that is,

$$\pi'(k_m)\left(1 + \frac{1}{1+r}\right) = h \tag{2.13a}$$

where r is the rate of discount. For the Cournot–Nash case the entrant's problem is to choose its output q^2 to maximize its profits net of capacity cost, given the second period output q_2^1 of the established firm. The established firm has to choose its output q_2^1 subject to its capacity constraint $q_2^1 \leq k^1$ and taking the entrant's output q^2 as given. Thus the incumbent's problem is:

$$\begin{array}{ll} \max_{q_2^1} & \left[p(q_2^1 + q^{2*})q_2^1 - c^1(q_2^1)\right] \\ \text{s.t.} & q_2^1 \leq k^1 \end{array}$$

while the entrant solves

$$\max_{q^2} \left[p(q_2^{1*} + q^2)q^2 - c^2(q^2) - hq^2\right]$$

where $(q_2^{1*} + q^{2*})$ is the capacity constrained Cournot–Nash equilibrium point. Now consider the capacity constrained post entry game. Given capacity k^1 the incumbent firm's reaction is truncated at k^1. This implies that unless the entry blocking output Q^1 is less than the short-run monopoly output, there is no possibility of the incumbent firm deterring entry. Entry will be deterred if the monopolist's optimal capacity level k_m in (2.13a) equals or exceeds Q^1. Thus if the incumbent follows a Cournot–Nash strategy, then entry is blocked if and only if

$$\pi'(Q^1)\left(1 + \frac{1}{1+r}\right) \geq h. \tag{2.13b}$$

Thus it is clear that the entry-deterring capacity level is chosen when the discount rates (r) are low or when the cost of capacity h is low relative to discounted marginal returns at the entry blocking output level Q^1. Also the established firm will produce at full capacity in each period. Spulber (1981) has further shown that an established firm faced with large-scale entry will only hold excess capacity before entry, when the firm is a Stackelberg leader with the Stackelberg output exceeding the firm's short-run monopoly output and the inequality (2.13b) holds strictly.

Note however that capacity investment may introduce economies of scale in the cost function and the relative uncertainty of potential entry may introduce risk averse behavior on the part of the established firm as shown before. In the general case when there are n firms $(n > 2)$, the entry and exit process can be modeled in terms of the fringe around the dominant firm. Novshek (1980) has discussed Cournot–Nash equilibrium with free entry assumption, when firms have size-specific average cost functions. Thus the α-size firm for each α corresponding to the long-run average cost AC is the firm AC_α defined by $AC_\alpha = AC(q/\alpha)$. The minimum efficient scale MES(AC) is then set equal to one for the basic AC function and hence α is used to generate the other average cost functions with different MES. Then it is shown that if firms are small relative to the overall market, then there is a Cournot–Nash equilibrium with free entry and any Cournot equilibrium with free entry is in the limit $n \to \infty$ approximately competitive. Thus the entry–exit dynamics of price and quantity adjustments would hold as discussed before.

Jovanovic and MacDonald (1989) have discussed competitive models of entry and exit behavior, when the cost functions depend on a technology parameter θ. New knowledge or "inventions" are assumed to emerge constantly in the industry and firms find a way to put it to used commercially, or to "innovate" in the Schumpeterian tradition. Those who succeed begin production or increase their share of industry output. Because innovation and progress lower unit costs for competitors, the less successful firms may find it to their advantage to exit. This is very similar to the Gaskins-type model. However, the success in new innovations is stochastic and this may cause some firms to lag behind as progress occurs.

2.3 Barriers to entry

Bain emphasized very strongly the hypothesis that firms in an industry can earn excess profits if they are protected by entry barriers. The barriers can take many forms. Salinger (1984) and others argued that Bain's hypothesis implies that the average profit rate depends on factors such as industry concentration, growth of past sales along with several measures of entry barriers such as advertising intensity, large size of MES and large strongholds of the leading firms. In the Schumpeterian tradition the role of new innovations which tend to lower average costs through higher MES and larger capital investment is important in setting up a potential entry barrier.

An alternative view is presented in the Chicago School's hypothesis that high profits are in fact a consequence of greater efficiency. Stigler (1958) proposed an interesting perspective on firm size: the "survivor technique". He stressed that only those firm sizes that survive the competition process in the long run determine the optimal size in a dynamic sense. To better understand the wide range of optimal firm sizes he performed an empirical inter-industry study and found that the advertising intensity and the capital/sales ratio were relatively insignificant in determining firm size, while the size of the plant and the relative volume of technical employment were highly significant. Sengupta

(2003) has recently applied this model to the computer industry over the last 15 years and found that R&D investment and net plant and machinery are very important in determining the optimal size and its change over time.

Two points are important here. One is the concept of growth efficiency vis-à-vis level efficiency, the former implying steady rates of growth of output for the growth efficient firms. The second aspect is that the presence of entry barriers would delay the entry process.

In recent econometric models developed by Geroski and Schwalback (1991) and Mata (1995) the rates of gross entry (ENT) and exit (EX) have been viewed as a simultaneous system, for example,

$$\begin{aligned} \text{ENT} &= \beta_1 = \beta_2\pi + \beta_3 \text{ BTE} \\ \text{EX} &= \delta_1 + \delta_2\pi + \delta_3 \text{ BTE} + \delta_4 \text{ ENT} \qquad (2.14) \\ \text{Hence} \quad \text{NETENT} &= \beta_1(1-\delta_4) - \delta_1 + (\beta_2(1-\delta_4) - \delta_2)\pi \\ &\quad + (\beta_3(1-\delta_4) - \delta_3) \text{ BTE}. \end{aligned}$$

Here BTE denotes barriers to entry variables and the signs of the coefficients are expected to be as follows: β_2 positive, β_3, δ_2, δ_3 negative and $1 > \delta_4 > 0$. The first equation predicts less entry with higher entry barriers, while the second equation says that *ceteris paribus* less exit will occur with higher profits and higher barriers. Mata (1995) estimated the system for Portuguese manufacturing over the period 1978–82. The BTE vector included product differentiation (PAT), economies of sale (LMES) and a measure of the degree of sunkness of machinery and equipment capital (SUNK). To control for different industry sizes, the logarithm of employment in the industry in 1982 (LSIZE) is included in the list of explanatory variables. Economies of scale were measured by the logarithm of the proxy of MES. Product differentiation increases the capital requirements barrier through advertising expenses and it was measured by the ratio of patents to production (PAT). By using the seemingly unrelated estimating technique the equations of (2.14) are estimated and corrected for heteroscedasticity by White's test. The results are in Table 2.1 with t values in parentheses.

The estimated results tend to confirm the hypotheses made about the determinants of entry at the 5 percent level of one sided t-test. All explanatory variables carry the expected signs and are significant at 5 percent level. Results for the exit equation also confirm that economies of scale and product differentiation are important in the explanation of exit levels, even after the effect of entry is taken into account.

Table 2.1 Determinants of entry and exit

	Entry	Exit	Exit	Net entry (*NETENT*)
Constant	−66.6	−26.3	−54.7	−11.9
	(−0.982)	(−2.122)	(−1.529)	(−0.348)
Profit	311.5	71.5	204.2	107.3
	(2.27)	(1.68)	(2.30)	(1.64)
Lsize	62.1	13.8	40.3	21.8
	(3.35)	(3.39)	(5.06)	(1.80)
Lmes	−44.8	−11.9	−31.0	−13.9
	(−3.21)	(−2.08)	(−3.69)	(−1.49)
Pat	−26.9	−7.2	−18.7	−8.2
	(−2.56)	(−2.12)	(−2.88)	(−1.45)
Sunk	−56.6	−3.6	−27.7	−28.4
	(−1.68)	(−0.95)	(−1.99)	(−1.33)
Entry	—	0.43	—	—
		(5.38)		
R^2/R^2	0.42/0.37	0.90/0.89	0.55/0.52	0.20/0.14

Note: The critical 5% value of corrected t value with 60 df is 1.671.

A second type of model based on the number of plants (N_t) is also developed by Mata as a partial adjustment type model

$$N_t - N_{t-1} = \lambda(N_t^* - N_{t-1}) \tag{2.15}$$

where N_t^* is the *long-run equilibrium* number of plants in the industry. The parameter λ measures the speed at which markets converge to equilibrium. If cost curves are standard U-shaped then N_t^* can be measured by the ratio size/MES$_t$, that is, the number of efficiency scaled plants (NES). Allowing for product differentiation (PAT) one can thus write

$$N_t^* = a_0 + a_1\text{NES}_t + a_2\text{PAT}_t. \tag{2.16}$$

Also Mata expresses λ as a function of sunk cost and profits as

$$\lambda = (\theta_1 + \theta_2\,\text{SUNK}_{t-1} + \theta_3\pi_{t-1})^2 \tag{2.17}$$

where π_t is profits at year t. Finally net entry NETENT $= N_t - N_{t-1}$ is expressed as

$$\text{NETENT} = (\theta_1 + \theta_2\,\text{SUNK}_{t-1} + \theta_3\pi_{t-1})^2[a_0 + a_1\text{NES}_t + a_2\text{PAT}_t - N_{t-1}]$$

and estimated by nonlinear least squares as follows:

θ_1	θ_2	θ_3	a_0	a_1	a_2	R^2
2.11	−0.69	2.06	−35.24	4.45	−19.10	0.92
(5.50)	(−4.25)	(1.92)	(−3.86)	(8.57)	(−5.22)	

when t values corrected for heteroscedasticity by White's test are in parenthesis and the critical value of one-sided t test at 5 percent level is 1.671 for $df = 60$. These estimated results show very clearly that net entry is restricted by high sunk cost and product differentiation acts as a significant barrier to entry.

A few comments on the economic implications of the model specified by (2.15) through (2.17) are in order. First of all, the entry concept used here is based on the actual number of firms joining the industry. But the decision to enter is always a subjective decision. Schumpeter emphasized the *subjectivity of knowledge* as the crucial factor and the dynamic role of entrepreneurs is to revolutionize the pattern of production by exploiting new inventions, for example, new commodities or new processes. Thus there is a stochastic element in the entry dynamics. If expected entry is denoted by $\overline{E}_t$, then it can be written as

$$\overline{E}_t = \alpha(\pi_{t-1} - r^*)$$

where $r^* = r^*(\text{BET, EFF})$ is long-run equilibrium rate of profits which depends on barrier to entry and efficiency measured by the cost advantage of the entering firm, for example, the gap between the industry average cost and the average cost of MES of the entering firm.

Subsequent to entering an industry a firm must decide whether to maintain its output (y_{jt}), expand, contract or exit. The probability of a firm remaining in business essentially depends on the firm's inherent size disadvantage and the probability of actually innovating in the Schumpeterian sense, that is,

$$\text{Prob}(y_{jt} > 0) = f(I_{jt}, c(y_{jt}) - c(y_j^*))$$

where $c(y_{jt})$ is the average cost of producing at a scale of output y_j for firm j and $c(y_j^*)$ is the average cost of producing the MES level of output and I_{jt} is the innovation process measured by R&D intensity.

An alternative way to measure potential entry is through the relative cost advantage. Thus if c_j is the unit cost of firm j and $\bar{c}$ the industry average cost, potential entry may be measured by $\dot{s}_j = ds_j/dt$ where s_j is the market share of output of firm j and we have the dynamics

$$\dot{s}_j/s_j = \lambda(\bar{c} - c_j), \quad j = 1, 2, \ldots, n. \tag{2.18}$$

It is assumed that firms exit, when their market shares are very low and the equation (2.18) states that a firm's market share (i.e., potential entry) grows if the firm is characterized by below-average unit costs (i.e., $c_j < \bar{c}$). The speed of selection or adjustment coefficient $\lambda (0 \leq \lambda \leq 1)$ may be constant or evolve over time. Also dynamic increasing returns or innovations efficiency may reduce c_j further, thus increasing the dominance of the more efficient firm. With high barriers to entry the selection process measured by λ may be slower. Thus the degree of concentration in an industry may have significant impact on the entry–exit dynamics.

2.4 Positive and negative feedback

Positive feedback essentially involves for dynamic economic systems increasing returns and multiple equilibria, some of which may be stable, some unstable. Agliardi (1998) has recently constructed dynamic economic models involving technological choice: whether to switch from the current to a new standard. These models study the economy as a complex system with self-reinforcement mechanisms. Self-reinforcement goes under different names such as increasing returns, cumulative causation and dynamic economies of scale. Young (1928) emphasized he role of increasing returns and dynamic economies of scale and Kaldor (1963) stressed the idea of cumulative causation, that is, success tends to breed success and failure also tends to be self-perpetuating thus causing unevenness in industrial development.

The theory of optimal investment decisions of firms shows that exploiting increasing returns is closely related to decision making under uncertainty and irreversibility. This is so because for a decision problem to be truly dynamic, there must be some uncertainty in the environment, for example, whether to invest now or wait for more information. Also most investment decisions are largely irreversible, because capital is firm-specific or industry-specific, that is, it cannot be used productively by a different firm or different industry, which makes its cost "sunk". The presence of irreversible investment expenditures generates *hysteresis*, that is, the cost and output effects persist even when the causes have disappeared. Dixit and Pindyck (1994) have shown that the ability to delay an irreversible investment has a profound effect on future output stream. This also causes firms to prefer more flexible decisions.

The above nonlinear effects are accentuated in the presence of indivisibilities, which add discontinuities and cause increasing returns to scale. Also, learning-by-doing and learning-by-using phenomena generate increasing returns to adoption of new innovation, because the more a technology is adopted, the larger the benefits of adoption, for example, network externalities.

Agliardi (1998) has discussed a dynamic stochastic model, where the firms have a choice problem: which technological standard to choose, when the two substitutable standards are denoted by zero and one. There are N firms in the industry and let $n(t)$ be the number of firms with standard 1 and $N - n$ with standard 0. Let $y = n/N$ be the proportion of N firms with standard 1. It is assumed that there are benefits from compatibility, that is, firms are able to exploit economies of scale in using a common supplier of a complementary good.

Following Agliardi we assume that $n(t)$ is a birth and death process with transition intensities $\lambda(y)$ and $\mu(y)$ for the transition $0 \to 1$ and $1 \to 0$ respectively. Agliardi has proved an important theorem which says that if $\lambda(z)$ and $\mu(z)$ are bounded and Lipschitz continuous, then for all $t \geq 0$ we have $\lim_{N\to\infty} E(y(t)) = z(t)$, where $z(t)$ satisfies the differential equation

$$dz/dt = (1 - z)\lambda(z) - z\mu(z), \quad z(0) = y(0). \tag{2.19}$$

The fixed points of (2.19) are the stationary solutions $\bar{z}$ (i.e., when $dz/dt = 0$) of the equation

$$\bar{z} = \lambda(\bar{z})/(\lambda(\bar{z}) + \mu(\bar{z})). \tag{2.20}$$

Two implications of the stationary solution are economically important. One is that under the assumption $\partial\mu(z)/\partial z < 0 < \partial\lambda(z)/\partial z$ the fixed point equation (2.20) has two sets of solutions: one set is asymptotically stable, the other unstable. Secondly, if we consider a one dimensional version of (2.20) and linearize it as a perturbation equation

$$\begin{aligned} du/dt &= [(1 - \bar{z})\partial\lambda(\bar{z})/\partial z - \lambda(\bar{z}) - \bar{z}\partial\mu(\bar{z})/\partial z - \mu(\bar{z})]u \\ &= -(1 - \partial F(\bar{z})/\partial z)(\lambda(\bar{z}) + \mu(\bar{z}))u \end{aligned}$$

where $F(\bar{z})$ is the right-hand side of (2.20), then it follows by the characteristic equation that $\bar{z}$ is stable if $\partial F(\bar{z})/\partial < 1$ and unstable if $\partial F(\bar{z})/\partial z > 1$.

Now consider a generalized version of the birth and death process model analyzed by Sengupta (1998), which assumes that the birth and death process intensities are nonlinear as follows:

$$\lambda = \alpha(k_2 - z)z, \qquad \mu = \beta(z - k), \quad k_1 < k_2. \tag{2.21}$$

Then $z(t)$ follows the Kolmogorov forward equation with mean $M(t)$ and variance $V(t)$:

$$dM(t)/dt = (\alpha + \beta)\left[\frac{\alpha k_2 + \beta k_1}{\alpha + \beta} M(t) - M^2(t) - V(t)\right]. \tag{2.22}$$

In the deterministic case this yields

$$dz/dt = (\alpha + \beta)[\theta z(t) - z^2(t)] \tag{2.23}$$

where $\theta = (\alpha k_2 + \beta k_1)/(\alpha + \beta)$.

This is a logistic growth process with a stationary solution $\bar{z} = 0$. Note some characteristics of this nonlinear model (2.22). First of all, the variance $V(t)$ of the process affect the rate of growth of $M(t)$ negatively. This is very similar to the Fisherian fitness model of replicator dynamics in genetic evolution. Second, the model above yields the equilibrium probability distribution derived from the appropriate Fokker–Planck equation, which shows that the relative fluctuations measured by the coefficient of variation $CV = \sigma/M$, $\sigma = \sqrt{V}$ may increase considerably in some domain. Furthermore, if we consider a two sector generalization of the above model as in Sengupta (1998), the interaction between the sectors may generate additional sources of instability. Here the two sectors may be the two groups of firms one with standard zero technology and the other with standard one. Alternatively they may represent one group of firms with sustained dynamic efficiency and the other with no such sustained efficiency.

The existence of unstable points of positive feedback systems may be easily compared to the adjustment processes we developed before in the selection mechanism. For example, if the quantity of output if q sold at price p and the average cost of production given by a smooth function $c(q)$, the adjustment mechanism in a competitive framework may be of the nonlinear form

$$\dot{q} = a[p - c(q)], \qquad \dot{p} = b[D(p) - q] \tag{2.24}$$

where a, b are positive constants and $D(p)$ is the demand function. Let $p^* = c(q^*)$ and $D(p^*) = q^*$ be the equilibrium solution. Then by linearizing the above system around the equilibrium point one may derive the system

$$\begin{pmatrix} \dot{q} \\ \dot{p} \end{pmatrix} = \begin{bmatrix} -ac' & a \\ -b & bD' \end{bmatrix} \begin{pmatrix} q \\ p \end{pmatrix}$$

where $c' = dc(q)/dq$ and $D' = dD/dp$. The characteristic equation in the eigenvalue θ is

$$\theta^2 - \theta(ac' - bD') + ab(1 - c'D') = 0.$$

It is clear that if $c'D > 1$ then there are two real solutions in θ, one positive and one negative. In this case the equilibrium point (q^*, p^*) is a saddle point, that is, one stable manifold along which motion is purely toward (q^*, p^*) and an unstable manifold along which motion is away from (q^*, p^*). Thus the industry moves toward one of two regimes. In one (the convergent case) outputs and profits rise and prices fall. In the other, profits fall and prices rise.

We have so far discussed the implications of positive feedback for the evolution of industry. If some firms have positive feedback and others have negative feedback (i.e., diminishing returns and dynamic diseconomies), then the interaction between these two groups of firms leads to a dominance of the positive feedback firms. Stigler's survivor technique emphasized that very strongly.

Consider for example two firm groups with outputs $y_i (i = 1, 2)$ growing exponentially

$$y_1(t) = y_{10} \exp(\lambda, t), \quad y_2(t) = y_{20} \exp(\lambda_2 t) \tag{2.25}$$

where λ_1, λ_2 are the net growth rates, where λ_i may represent the difference of the birth rate and the death rate intensities. If $\lambda_1 > \lambda_2 > 0$ due to the fact that the rate of Schumpeterian innovation or the dynamic efficiency is higher for the first group, then the growth rate of the mixture $y(t) = y_1(t) + y_2(t)$ follows the dynamic process

$$d \ln \dot{y}(t)/dt = (d \ln y/dt)[\lambda_1 - d \ln y(t)/dt].$$

Clearly as $t \to \infty$ the total output $y(t)$ tends to λ_1, which is the relative growth rate of the more efficient group. The average gain in efficiency for the industry defined as

$$E(t) = (\dot{y}(t)/y(t)) - \lambda_2$$

then follows the time-path

$$E(t) = s(\theta e^{-st} + 1)^{-1}$$

where $\theta = y_{20}/y_{10}$ and $s = \lambda_1 - \lambda_2 > 0$. Thus $E(t)$ tends to s as $t \to \infty$. Here the parameter s is the efficiency advantage of the higher efficiency type over the lower efficiency type.

2.4.1 Survival of the fittest

The survivor technique initially proposed by Stigler provides a market selection process of empirically identifying the minimum efficient scale (MES) of a firm in an industry. Thus it applies in an approximate sense the principle of survival of the fittest, where fittest represents the most dynamically efficient group of firms. It provides a culmination of the entry–exit process, till the long-run equilibrium is reached with the surviving firms. We consider here some nonparametric extensions of the survivor technique, which are based on estimating a cost frontier and then characterizing an intertemporal growth path for the efficient firms in the industry. A dynamic process for entry and exit of firms can be suitably modeled in this framework, where positive (negative) profits invite new entry (exit).

The survivor technique is a method of estimating economies of scale in long-run costs. Stigler (1958) first proposed this method and applied it to the US steel industry. His method was to observe an industry over time, classify the firms in the industry by size (measured as a percentage of total industry capacity or output) and then arrive at a conclusion regarding cost-efficiency based on the relative growth or decline of these size categories. Following this method Rogers (1993) classified US steel mills into four size categories by their annual capacity (1–1.49 mt, 1.5–4.5 mt, 4.5–7.5 mt and over 7.5 mt) in 1976 and 1987. He found that the size group 4.5 to 7.5 mt increased its market share to the highest degree from 19.3% in 1976 to 28.1% in 1987. Hence he concluded that this size group is the minimum optimal size with the most efficient scale.

The survivor technique hypothesizes that only the efficient size plants will survive in competition with other plants. Furthermore, if one plant size was more efficient than the others, its share of total market would increase. This technique may be viewed as an application of statistical cost analysis, where the most efficient scale is identified by time series observations.

Although very practical and convenient the technique has several basic weaknesses. First, it cannot determine the exact nature and size of the advantage of the efficient plant. Since the technique does not distinguish between a cost function and a cost frontier, it fails to identify the sources of cost efficiency. Second, the flexibility of the long-run cost function is an important determinant of the scale economies in cost and this can only be specified by a nonlinear cost frontier and the survivor technique fails to identify this feature of cost-efficiency. Finally, the technique is of little help in measuring costs for planning purposes. It merely tells us which company size appears to be more efficient; it says nothing about relative costs. Also changing technology and the change in distribution of firms in the industry may cause distortions so that firms of certain size are favored over others.

Consider a sample of N firms each producing a single output (y_j) with m inputs. Let $C_j = \sum_i q_i x_{ij}$ be the total cost of inputs where q_i is the unit cost of input x_i and c_j be the average cost $c_j = C_j/y_j$ for $j = 1, 2, \ldots, N$. To test the relative cost efficiency of firm h, that is, if it is on the convex hull of the cost output space we set up the linear programming (LP) model

$$\begin{aligned} &\text{Min} \quad \theta \\ &\text{s.t.} \quad \sum_{j=1}^{N} \lambda_j c_j \leq \theta c_h; \quad \sum_{j=1}^{N} \lambda_j y_j \geq y_h \\ &\qquad \sum_{j=1}^{N} \lambda_j = 1, \ \ \lambda_j \geq 0; \quad j = 1, 2, \ldots, N. \end{aligned} \tag{2.26}$$

This is usually called the input-oriented efficiency model in the DEA literature, see for example, Charnes *et al.* (1994) and Sengupta (2003). Let $\lambda^* = (\lambda_j^*)$ and θ^* be the optimal solutions of the LP model (1) with all slack variables zero. Then the reference unit or firm h is cost efficient, that is, on the cost frontier if $\theta^* = 1.0$. If however θ^* is positive but less than one, then it is not cost efficient at the 100% level, since it incurs excess input costs measured by $(1 - \theta^*)c_h$. On using the Lagrangean function

$$L = -\theta + \beta\left(\theta c_h - \sum_j \lambda_j c_j\right) + \alpha\left(\sum_j \lambda_j y_j - y_h\right) + \beta_0\left(\sum_j \lambda_j - 1\right)$$

the duality theorem implies the following cost frontier for firm j, when it is cost efficient:

$$c_j = \gamma_0 + \gamma_1 y_j; \quad \gamma_0 = \beta_0/\beta, \quad \gamma_1 = \alpha/\beta. \tag{2.27}$$

Several generalizations of this model are in order. First of all, consider the long-run cost implications by introducing a new capital output ratio variable k_j through the constraint

$$\sum_j \lambda_j k_j \leq \theta k_h.$$

The introduction of the capital variable is helpful for two reasons. One is that it can show the learning curve effects. The other is that it may represent R&D type innovations. The cost frontier for firm j appears as

$$c_j = \gamma_0 + \gamma_1 y_j - \gamma_2 k_j; \qquad \gamma_2 = b/\beta > 0 \tag{2.28}$$

where b is the Lagrange multiplier for the capital constraint above. Now average cost for the efficient firm j tends to fall as investment in capital per unit of output increases. Note that k_j may represent R&D investment in new technology or a proxy for cumulative investment specifying learning-by-doing. Note also that the case of multiple outputs is easily handled through standard methods in DEA approach, where output y_j is viewed as a vector.

A second type of generalization introduces more flexible cost functions. To see this, replace average cost c_j by total cost C_j in model (2.26) and add the second order constraint on output as

$$\sum_j \lambda_j y_j^2 \geq y_h^2. \tag{2.29}$$

Then if firm h is cost efficient with $\theta^* = 1.0$ we obtain the quadratic cost frontier

$$C_h = \gamma_0 + \gamma_1 y_h + \gamma_2 y_h^2; \quad \gamma_0 = \beta_0/\beta, \ \ \gamma_1 = \alpha/\beta, \ \ \gamma_2 = a/\beta$$

where a is the Lagrange multiplier for the quadratic constraint (2.29). This quadratic cost function model specifies variable returns to scale in a more direct way. Here we test the relative cost-efficiency of a firm h in relation to other firms in the industry by specifying the convex hull in the cost output space spanned by the N firms. The following linear programming (LP) model provides the test:

$$\begin{aligned} \text{Min} \quad & \varepsilon_h = C_h - C_h^* \\ \text{s.t.} \quad & \gamma_0 + \gamma_1 y_j + \gamma_2 y_j^2 \leq C_j, \quad j = 1, 2, \ldots, N \end{aligned}$$

where the optimal cost is C_j^* which is specified by $C_j^* = \gamma_0 + \gamma_1 y_j + \gamma_2 y_j^2$ such that $C_j \geq C_j^*$ and $\varepsilon_j = C_j - C_j^*$ is the relative inefficiency. The dual form of this model can be used to derive the transformed model as follows:

$$\begin{aligned} \text{Min} \quad & \theta \\ \text{s.t.} \quad & \sum_{j=1}^{N} \lambda_j C_j \leq \theta C_h; \quad \sum_j y_j \lambda_j \geq y_h \\ & \sum_j y_j^2 \lambda_j \geq y_h^2; \quad \Sigma \lambda_j = 1; \quad \lambda_j \geq 0. \end{aligned}$$

Note that this provides the rationale for the quadratic constraint (2.29) above. The cost frontier for this model is then of the form

$$C_h = C_j^* = \gamma_0 + \gamma_1 y_h + \gamma_2 y_h^2$$

if firm h is on the frontier. The optimal average cost for this cost frontier is then

$$c_h = (\gamma_0/y_h) + \gamma_1 + \gamma_2 y_h. \tag{2.30a}$$

We may minimize this further to obtain the output y_h^* yielding the lowest optimal average cost, that is,

$$y_h^* = (\gamma_0/\gamma_2)^{1/2}; \quad c_h^* = 2(\gamma_0\gamma_2)^{1/2} + \gamma_1. \tag{2.30b}$$

Clearly if the cost frontier can be decomposed into short-run and long-run components under time series data, then a more generalized DEA model may be formalized as follows:

$$\begin{aligned}
\text{Min} \quad & \theta(t) \\
\text{s.t.} \quad & \sum_{j=1}^{N} \Delta C_j(t)\lambda_j(t) \leq \theta(t)\Delta C_h(t) \\
& \sum_j \Delta y_j \lambda_j(t) \geq \Delta y_h(t) \\
& \sum_j C_j(t-1)\lambda_j(t) \leq C_h(t-1) \\
& \sum_j y_j(t-1)\lambda_j(t) \geq y_h(t-1); \quad \sum_j y_j^2(t-1)\lambda_j(t) \geq y_h^2(t-1) \\
& \sum_j \lambda_j(t) = 1; \quad \lambda_j(t) \geq 0; \quad j = 1, 2, \ldots, N.
\end{aligned} \tag{2.31}$$

The optimal cost frontier then reduces to

$$\begin{aligned}
\Delta C_j(t) = {} & \delta \Delta y_j(t) - (b/\beta)[C_j(t-1) - \gamma_1 y_j(t-1) \\
& - \gamma_0 - \gamma_2 y_j^2(t-1)]
\end{aligned} \tag{2.32}$$

in terms of change in costs, where the Lagrangean function is

$$\begin{aligned}
L = {} & -\theta(t) + \beta\left(\theta(t)\Delta C_h(t) - \sum_j \Delta C_j(t)\lambda_j(t)\right) \\
& + \alpha\left(\sum_j \Delta y_j(t) - \Delta\, y_h(t)\right)
\end{aligned}$$

$$+\,b\Big(C_h(t-1)-\sum_j C_j(t-1)\lambda_j(t)\Big)$$

$$+\,a_1\Big(\sum_j y_j(t-1)\lambda_j(t)-y_h^{(}t-1)\Big)$$

$$+\,a_2\Big(\sum_j y_j^2(t-1)\lambda_j(t)-y_h^2(t-1)\Big)$$

$$+\,\beta_0\Big(\sum_j \lambda_j(t)-1\Big)$$

and $\delta = \alpha/\beta$, $\gamma_0 = \beta_0/b$, $\gamma_1 = a_1/b, \gamma_2 = a_2/b$. If total costs and output are first difference stationary, then this cost frontier (2.32) is the error correction model (ECM), where both least squares and least absolute value methods of estimation are directly applicable, whereas such applications are statistically invalid for ordinary regression of C_t on y_t, since each series is a random walk. The dynamic efficiency model (2.32) has some interesting interpretations. First of all, the frontier has two components: the short-run component when b/β is close to zero, that is, $\Delta C_j(t) = \delta \Delta y_j(t)$ and the long-run component when δ is close to zero, that is,

$$C_j(t-1) = \gamma_0 + \gamma_1 y_j(t-1) + \gamma_2 y_j^2(t-1), \text{ that is,}$$
$$\overline{C}_j = \gamma_0 + \gamma_1 \bar{y}_j + \gamma_2 \bar{y}_j^2 \text{ in the steady state.} \tag{2.33}$$

Note that the short-run marginal cost δ would differ from the long-run marginal cost given by $(\gamma_1 + 2\gamma_2\bar{y})$. Second, the long-run or steady state cost frontier (2.33) yields a steady state average cost $(\bar{c}_j)$ as:

$$\mathrm{AC}_j = \bar{c}_j = (\gamma_0/\bar{y}_j) + \gamma_1 + \gamma_2 \bar{y}_j.$$

Minimizing this average cost yields as before the most efficient scale of output as

$$\bar{y}_j^* = (\gamma_0/\gamma_2)^{1/2}; \quad \bar{c}_j^* = 2(\gamma_0\gamma_2)^{1/2} + \gamma_1. \tag{2.34a}$$

Let $N_1 < N$ firms be dynamically efficient in the long-run sense (2.33) for $j = 1, 2, \ldots, N_1$. Then we can arrange these firms in an ascending order:

$$\bar{c}_{(1)}^* \leq \bar{c}_{(2)}^* \leq \bar{c}_{(3)}^* \leq \cdots \leq \bar{c}_{(N_1)}^*. \tag{2.34b}$$

The associated output levels may be denoted by $\bar{y}^*_{(j)}$, $j = 1, 2, \ldots, N_1$. Clearly by the survivor technique the size class $(\bar{c}^*_{(1)}, \bar{y}^*_{(1)})$ would indicate the most efficient scale. This measure of the most efficient scale is more flexible and appropriate than that determined by the survivor technique. This is so because it is based on a fully specified convex cost frontier for each firm that is dynamically efficient in the sense of (2.30b). Unlike the survivor technique it follows a two-step non-parametric procedure for identifying the most efficient scale. The first step is to identify the subset of N_1 firms each of which is dynamically cost efficient. The second step then identifies the most efficient scale in terms of the value $\bar{c}^*_{(1)}$ of lowest average cost at an output level $\bar{y}^*_{(1)}$. This two-step method is also better than the standard scale efficiency measure used in traditional DEA models, because it identifies the optimal *output size* at which the average cost frontier reaches its minimum. In the traditional DEA model the scale efficiency (SE) is measured by

$$\mathrm{SE} = \theta^*_{\mathrm{CRS}}(t)/\theta^*_{\mathrm{VRS}}(t) \tag{2.35a}$$

where $\theta^*_{\mathrm{CRS}}(t)$ is the optimal value of θ in LP model (1) with the constraint $\sum \lambda_j = 1$ dropped implying constant returns to scale (CRS) and $\theta^*_{\mathrm{VRS}}(t)$ is the value of θ^* when the constraint $\sum \lambda_j = 1$ is not dropped. The optimal output size is not explicit here.

This two-step method of identifying the most efficient scale is however limited by two conditions. One is the sample size N. If the sample size is not adequately representative of the whole industry, then it may yield a biased measure. Secondly, the value of the parameter γ_2 associated with the quadratic component of the total cost frontier is very critical here. The higher the value of $\gamma_2 > 0$, the higher is the output level $\bar{y}^*_j$ associated with $\bar{c}^*_j$ for an efficient firm j. If γ_2 is dropped, the minimal average costs $\bar{c}^*_{(j)}$ can still be determined as in (2.34a), so that $\bar{c}^*_{(1)}$ defines the most efficient scale among the N_1 efficient firms.

The process of entry and exit of firms or the increase and decrease of market share can be easily modeled here in terms of the competitive industry comprising the firms. For this purpose assume that firms are identified by their cost structures, where the output y_j of each firm j belongs to one of K possible types of cost structures. We have to select the optimal scale of output y_j of each firm which minimizes total costs (C) for the whole industry, that is,

$$\text{Min} \quad C = \sum_{j=1}^{K} n_j C_j(y_j)$$

$$\text{s.t.} \quad \sum_{j=1}^{K} n_j y_j \geq D; \quad y_j \geq 0; \quad n_j \geq 0.$$

Here $C_j(y_j) = \gamma_0 + \gamma_1 y_j + \gamma_2 y_j^2$ as before. For a given market demand D, let $\hat{y} = (y_j), \hat{n} = (n_j)$ be the vectors of optimal output and the number of firms in different cost structures and let $\hat{p}$ be the optimal value of the Lagrange multiplier. We may interpret $\hat{p}$ as the market clearing price which equates total demand and supply. Let $C = C(\hat{n}, D)$ be the total cost function for the industry as a whole. It can be easily shown in terms of the Hessian matrix that C is strictly convex in vector $\hat{n}$. Hence a minimizing value $\hat{n}^*$ exists such that $C(\hat{n}) \geq C(\hat{n}^*)$, where $\hat{n}^*$ satisfies the conditions

$$\phi_j(\hat{n}^*) \geq 0 \text{ and } \hat{n}_j \phi_j(\hat{n}^*) = 0, \quad \text{all } j = 1, 2, \ldots, K$$

$$\text{where} \quad \phi_j(\hat{n}^*) = [\hat{y}_j(C_j'(y_j) - AC_j)] = [\hat{y}_j \hat{p}(\hat{n}^*, D) - C_j(y_j)].$$

Since price (p) equals marginal cost $C_j'(y_j) = \text{MC}_j$ for each firm j. The dynamics of entry (or increased market share) and exit (or decreased market share) of firms may then be modeled as

$$\frac{dn_j}{dt} = \begin{cases} h_j(\text{MC}_j - \text{AC}_j); & h_j > 0 \text{ and } n_j > 0 \\ \max(0, h_j(\text{MC}_j - \text{AC}_j)); & n_j = 0 \end{cases}$$

where h_j is the adjustment coefficient and the condition $\text{MC}_j = \text{AC}_j$ specifies the equilibrium number of firms. Here positive excess profits $\hat{y}_j \hat{p} - C_j(y_j) > 0$ invite new entry (or increased market share for efficient firms) and negative expected profits force old firms to exit (or reduce their market share).

The survivor technique has two basic problems whenever we analyze the long-run trend of growth, survival and relative decay of firms in the industry. First of all, the long-run cost function of a firm depends on technological innovation and hence the growth of total factor productivity (TFP) is as important as scale economies with a given technology. Even the nonparametric DEA models (2.30a) or (2.30b) do not specifically introduce TFP growth as a basic factor which may reduce average costs.

The implications of TFP growth for firms in an industry can be directly analyzed through a dynamic production frontier approach using the DEA framework, which is based on a Malmquist index measure. This has been analyzed in some detail by others in terms of a production frontier approach. In our case we need to develop a cost-efficiency based approach which serves as the basis of generalized DEA models. Secondly, the survivor technique does not characterize the time path through which an efficient (inefficient) firm grows (declines) over time, when its cost frontier exhibits significant economies (diseconomies) of scale and substantial TFP growth (decline). We need to develop a dynamic intertemporal path for the growth of a dynamically efficient firm.

Empirically the dynamic role of technological innovation through R&D and learning-by-doing has been emphatically established by the econometric studies in recent times for the high-tech industries such as microelectronics, semiconductors, computers and telecommunications. Thus Norsworthy and Jang (1992) in their econometric measurement of technological change (progress) in these industries over the last decade noted the high degree of cost-efficiency due to learning-by-doing and R&D investment. Thus they estimated a translog total cost function for the US computer industry (SIC codes 3570, 3571) over the 23 year period 1959–81 and found the TFP growth to be 26%, whereas it is about 0.9% for manufacturing. Also they found significant degrees of increasing returns to scale – for the computer industry, for example, 2.88 on the average. Similar is the trend for other high-tech industries like semiconductors and microelectronics. Recently the empirical study by Jorgenson and Stiroh (2000) also found substantial TFP growth in the computer and related industries. High productivity growth leads to falling unit costs and prices and average computer prices, according to their estimate, fell by 18% from 1960 to 1995 and by 27.6% per year over 1995–98.

Norsworthy and Jang used a partial form of the translog cost function by omitting the input prices as:

$$\ln \text{TC} = b_0 + b_1 \ln y + b_2 t.$$

On taking the time derivative one obtains

$$\Delta\text{TC}/\text{TC} = b_1(\Delta y/y) + b_2. \tag{2.35b}$$

Here time t is used as a proxy for technological change implying a downward shift of the cost function, that is, technological progress when $b_2 < 0$. The inverse of the parameter b_1 is the degree of returns to scale and a negative (positive) value of the parameter b_2 indicates technological progress (regress). A quadratic form of the cost function, that is, quadratic in $\ln y$ would imply:

$$\Delta TC/TC = (b_1 + 2b_3 \ln y)(\Delta y/y) + b_2. \tag{2.35c}$$

Note that the cost frontier based on (2.35b) or (2.35c) would involve growth in costs due to growth in output and technological change, hence we can specify *growth efficiency* frontier, which is distinct from the *level efficiency* involving cost and output levels only. For example a level efficiency DEA model can be set up as

$$\begin{aligned} &\text{Min} \quad \theta \\ &\text{s.t.} \quad \sum_{j=1}^{N} \hat{C}_j \lambda_j \leq \theta \hat{C}_h; \quad \sum_j \hat{y}_j \lambda_j \geq \hat{y}_h \\ &\qquad \sum_{j=1}^{N} \lambda_j \leq 1; \quad t = 1, 2, \ldots, T; \lambda_j \geq 0. \end{aligned}$$

Here $\hat{C}_j = \ln TC_j$, $\hat{y}_j = \ln y_j$. If firm j is level efficient, then its cost frontier is

$$\hat{C}_j = (\alpha/\beta)\hat{y}_j - bt$$

where the Lagrangean is $L = -\theta + \beta(\theta\hat{C}_h - \sum_j \hat{C}_j\lambda_j) + \alpha(\sum_j \bar{y}_j\lambda_j - \bar{y}_h) + b(t - \sum_j \lambda_j t)$.

By imposing the condition $\sum \lambda_j t = t$ the optimal value of the Lagrange multiplier b may be made free in sign, so that technological regress can also be measured, that is, $b < 0$ (regress), and $b > 0$ (progress).

In case of growth efficiency frontier, for example, (2.35b) the DEA model takes the following form:

$$\begin{aligned} &\text{Min} \quad \phi(t) \\ &\text{s.t.} \quad \sum_j \tilde{C}_j(t)\lambda_j(t) \leq \phi(t)\tilde{C}_h(t) \\ &\qquad \sum \tilde{y}_j(t)\lambda_j(t) \geq \tilde{y}_h(t) \\ &\qquad \sum_j \lambda_j(t) \leq 1, \quad j = 1, 2, \ldots, N; \quad t = 1, 2, \ldots, T \end{aligned} \tag{2.36}$$

where $\tilde{C}_j(t) = \Delta C_j(t)/C_j(t)$, $\tilde{y}_j(t) = \Delta y_j(t)/y_j(t)$.

The nonparametric dynamic cost frontier for firm j now takes the form:

$$\begin{aligned} \Delta C_j(t)/C_j(t) &= (\alpha/\beta)(\Delta y_j(t)/y_j(t)) - b \\ &= b_1(\Delta y_j/y_j(t)) + b_2 \end{aligned} \tag{2.37}$$

where $b_1 = \alpha/\beta$ and $b_2 = -b$. For $b_2 < 0$ we have technological progress and $b_2 > 0$ represents technological regress. Note that if over time TC_j and y_j follow a geometric random walk process over time, so that the first differences of ln TC and ln y are stationary, then the growth efficiency model (2.36) in an estimation framework has parameters that are structurally stable. Furthermore the growth efficiency model characterizes the intertemporal growth frontier: $\{\phi^*(t); \lambda^*(t); \hat{C}_j(t)\}$ as time t evolves. If it converges to a steady state, then the steady state can be interpreted as a long-run growth frontier, as Solow (1957, 1997) interpreted technological progress in his long-run growth model.

In a recent study Sengupta (2003) applied the level efficiency and growth efficiency models in the computer industry for 20 firms over a 12-year period 1987–98. He found that all firms (10% of the total) which are growth efficient are also level efficient each year, but the obverse is not true, that is, all level efficient firms are not growth efficient. The implication for the survivor principle is clear. One has to concentrate on the core firms which are both level and growth efficient and then determine the sources of this efficiency in terms of scale efficiency and technological progress in terms of the nonparametric frontier (2.37).

The dynamics of the cost frontier (2.35) has two intertemporal aspects that are important in any industry which is highly competitive and oriented to modern technology such as telecommunications, microelectronics and computers. One is the entry and exit pattern of firms in an industry, which may be due to unequal rates of productivity growth and efficiency. The second is the dynamics of the relative efficiency distribution across firms in an industry, where the competitive selection process selects the efficient firms by the principle of survival of the fittest. The first aspect has been studied empirically by Lansbury and Mayes (1996) over UK manufacturing data for 1980 and 1990. They found that the competitive process involves not just the development of existing firms but new entrants who challenge the incumbents. The productivity and efficiency of most new entrants is higher than that of the sample as a whole in most years from 1980 to 1990. Similarly in most years most exits had a lower productivity than the sample as a whole.

Note however that productivity is a more basic concept than cost-efficiency, since it includes such factors as organizational efficiency, core competence and the quality of human capital employed by firms. Here cost-efficiency is used as a partial equilibrium measure, because it captures the adjustment process of a competitive market under free entry and free exit conditions. Hence we have used here an efficiency-based entry–exit model.

The second aspect has been studied theoretically by Jovanovic (1982), who views two groups of firms in an industry, where the competitive process provides a random selection mechanism. He assumes that the first group of firms is more efficient than the second group at all levels of output. The average cost function of the inefficient firms is $c(x_t)\varepsilon_t$, where ε_t is assumed to be a random variable independent across two groups of firms, where larger ε_t's generate more inefficiency. The market selection process in this model starts from the initial inequality in efficiency, which generates inequality in firm sizes. This means that large firms earn higher rents and profits. But the marginal firms, that is, small firms do not earn rent; there is no reason to expect a positive relation between concentration and the profits of these firms.

We incorporate these aspects in a dynamic model of growth of the efficient firms, when the exit rates depend on market price and inefficiency. We consider first a deterministic intertemporal model of growth of firms that are dynamically efficient, when the inefficient firms feel the pressure to exit. Let y_t and x_t be the outputs of two representative firms, one efficient and the other inefficient, when efficiency is viewed in terms of equation (2.34b). Let c_0 and c be the respective minimum average costs for the two firms such that $c = \varepsilon c_0$, $\varepsilon \geq 1$, where ε is the average rate of inefficiency. Let $E = -\dot{x} = -\mathrm{d}x/\mathrm{d}t$ be the rate of exit, where

$$\dot{E} = h(\varepsilon, p_t) = k_1\varepsilon - k_2 p_t \tag{2.38}$$

$$\delta h/\delta\varepsilon > 0, \quad \delta h/\delta p_t < 0; \quad k_1, k_2 \geq 0.$$

The average industry price is $p_t = a + b_1 E_t - b_2 y_t$. The exit behavior in equation (2.38) implies that higher inefficiency in the form of larger ε would induce greater exit rate, whereas higher average price p_t would tend to induce higher entry or lower exit rates. The efficient firm then solves the dynamic profit maximization problem

$$\begin{aligned} \text{Max} \quad & J = \int_0^\infty e^{-rt}(p_t - c_0)\,\mathrm{d}t \\ \text{s.t.} \quad & (2.38) \end{aligned} \tag{2.39}$$

Pontryagin's maximum principle may be applied to solve this dynamic model. Let the Hamiltonian function be

$$H = e^{-rt}[(p_t - c_0)y + \pi_t(k_1\varepsilon - k_2p_t)].$$

The optimal output strategy of the efficient firm must then satisfy the following necessary conditions along the optimal trajectory

$$\begin{aligned} \dot{E}_t &= k_1\varepsilon - k_2(a + b_1E_t - b_2y_t) \\ \dot{\pi}_t &= -\delta H/\delta E_t \\ \lim_{t\to\infty} \pi_t &= 0 \quad \text{and} \quad \delta H/\delta y_t = 0. \end{aligned} \tag{2.40}$$

Since the instantaneous profit function is a smooth strictly concave function of the efficient firm's output, it is easily shown that the above necessary conditions are also sufficient. On applying (2.40) one obtains the following two linear differential equations characterizing the optimal trajectory:

$$\begin{aligned} \dot{\pi}_t &= \alpha_2\pi_t - (b_1^2E_t/2b_2) - A_1 \\ \dot{E}_t &= \alpha_1E_t + k_2^2\pi_t/2 + A_2 \end{aligned} \tag{2.41}$$

where

$$\alpha_1 = \frac{b_1k_2}{2b_2} - k_2, \quad \alpha_2 = r + b_1k_2 - \frac{b_1k_2}{2}$$

$$A_1 = b_1a_1/(2b_2), \quad A_2 = k_1\varepsilon - k_2a + k_2(a - c_0)/(2b_2).$$

On eliminating the adjoint variable π_t we obtain the reduced form

$$\begin{aligned} \dot{E}_t &= -k_2E_t + k_2y_t + A_2 - k_2a_1/(2b_2) \\ \dot{y}_t &= \alpha_3E_t + \alpha_4y_t + A_3 \\ \text{where} \quad \alpha_3 &= (2b_1b_2\alpha_1 - b_2^3k_2 - 2\alpha_2b_1b_2 - b_1^2k_2)\,(4b_2^2)^{-1} \\ \alpha_4 &= (2b_2)^{-1}(2\alpha_2b_2 + b_1k_2) \\ A_3 &= (4b_2^2)^{-1}(b_1A_2 - k_1b_2A_1 + b_1a_1k_2 - 2a_1\alpha_2b_2). \end{aligned} \tag{2.42}$$

Several implications follow from these optimal trajectories. First of all, if y_t rises, then it increases the exit rate of inefficient firms. Also if $\dot{y}_t$ rises, then it increases the exit rate. This is more clear when r is zero and $b_1 = b_2 = b$, for then we have $\alpha_3 = 2k_2b^3$. Second, the higher the inefficiency index ε, the greater is the exit rate, so that the efficient firm's growth is efficiency driven. Finally, the higher b_2 implies lower A_2 and lower A_2 in turn generates lower prices for both efficient and inefficient firms. But the pressure of lower prices affects the inefficient firms more adversely than the efficient firms, since the latter can reduce their optimal average costs over time. This helps to explain the long-run dominance in market share of the efficient firm, as the survivor principle implies.

Next we consider the inefficient firm's behavior under random inefficiency parameter ε_t, where ε_t is assumed to be independently distributed as a half-normal variable with a nonnegative domain with mean $\bar{\varepsilon}$ and variance v. It is assumed that the inefficient firm producing x_t is risk averse and it maximizes the present value of the expected utility of profits as

$$\begin{aligned} \text{Max} \quad & J = \int_0^\infty e^{-rt}[px_t - c_0 x_t \bar{\varepsilon} - rc_0^2 x_t^2 v]\, \mathrm{d}t \\ \text{s.t.} \quad & \dot{y}_t = k(p - c_0), \quad p = a - b_1 x_t - b_2 y_t. \end{aligned}$$

Here c_0 is the minimum average cost of the efficient firm as before, r is the positive rare of risk aversion, and the condition $p > c_0$ is sufficient for successful entry of the new firm. Since the average market price $p_t = p$ is a market clearing price, any decline in x_t is offset by an increase in y_t, otherwise market prices may fall as y_t increases. This model can be solved as before by forming the Hamiltonian

$$H = e^{-rt}[px_t - c_0\bar{\varepsilon}x_t - rc_0^2x_t^2v + k\pi(p - c_0)].$$

The necessary conditions for an optimal path are

$$\delta H/\delta x_t = 0; \quad \dot{\pi}_t = -\delta H/\delta y_t; \quad \lim_{t\to\infty} \pi_t = 0$$

which are also sufficient here due to the concavity of the objective function with respect to x_t. Note that the short-run profit (or loss) function yields the result

$$x_t^* = (2b_1 + 2rc_0^2 v)^{-1}(a - c_0\bar{\varepsilon} - b_2 y)$$

which implies that the optimal output x_t^* will fall as $\bar{\varepsilon}$ or v rises. Likewise the impact of higher y is to reduce x_t^*. This process will be accentuated in the long run when the efficient firm's market entry increases in the form of $\dot{y} > 0$.

The instantaneous condition $\partial H/\partial x_t = 0$ at each t on the optimal trajectory also implies that

$$x_t^{**} = x_t^* - kb_1(2b_1 + 2rc_0^2 v)^{-1}\pi_t^{**}$$

where double asterisks indicate the optimal trajectory. Thus if $\pi_t^{**} > 0$, then $x_t^{**} < x_t^*$, that is, the inefficient firm declines more over time as the market share of the efficient firm increases. Furthermore the short term optimal output x_t^* is a strictly convex function of the variance of ε, it can be shown that the equilibrium price will tend to fall with higher variance. This result agrees with that of Jovanovic, although it is derived here through an intertemporal optimization process.

References

Agliardi, E. (1998): *Positive Feedback Economies*. St. Martin's Press, New York.

Bain, J.S. (1956): *Barriers to New Competition*. Harvard University Press, Cambridge.

Charnes, A., Cooper, W.W., Lewin, A. and Seiford, L. (1994): *Data Envelopment Analysis*. Kluwer Academic Publishers, Dordrecht.

Dixit, A. and Pindyck, R.S. (1994): *Investment Under Uncertainty*. Princeton University Press, Princeton.

Dreze, J. and Sheshinski, E. (1984): On industry equilibrium under uncertainty. *Journal of Economic Theory* 33, 88–97.

Gaskins, D.W. (1971): Dynamic limit pricing: optimal pricing under threat of entry. *Journal of Economic Theory* 3, 306–22.

Geroski, P. and Schwalbach J. (eds) (1991): *Entry and Market Contestability: An International Comparison*. Basil Blackwell, Oxford.

Jorgenson, D.W. and Stiroh, K.J. (2000): Raising the speed limit: US economic growth in the information age. In: Brainard, W. and Perry, G. (eds) Brookings Papers on Economic Activity. Brookings Institution, Washington, D.C.

Jovanovic, B. (1982): Selection and the evolution of industry. *Econometrica* 50, 649–70.

Jovanovic, B. and MacDonald, G. (1989): Entry and exit in perfect competition. Working paper No. 190, Department of Economics, New York University, New York.

Kaldor, N. (1963): Capital accumulation and economic growth. In: Lutz, F. and Hague, D.C. (eds), *Proceedings of the International Economic Association*. Macmillan, London.

Klepper, S. (1996): Exit, entry, growth and innovation over the product life-cycle. *American Economic Review* 86, 562–83.

Lansbury, M. and Mayes, D. (1996): Entry, exit, ownership and the growth of productivity. In: Mayes, D.G. (ed.), *Sources of Productivity Growth*. Cambridge University Press, Cambridge.

Mata, J. (1995): Sunk costs and the dynamics of entry in portuguese manufacturing. In: Witteloostuijn, A. (ed.), *Market Evolution: Competition and Cooperation*. Kluwer Academic Publishers, Dordrecht.

Mazzucato, M. (2000): *Firm Size, Innovation and Market Structure*. Edward Elgar, Cheltenham.

Norsworthy, J.R. and Jang, S.L. (1992): *Empirical Measurement and Analysis of Productivity and Technical Change*. North Holland, Amsterdam.

Novshek, W. (1980): Cournot equilibrium with free entry. *Review of Economic Studies* 47, 473–86.

Orr, D. (1974): The determinants of entry: a study of the Canadian manufacturing industries. *Review of Economics and Statistics* 56, 94–106.

Rees, R.D. (1973): Optimum plant size in UK industries: some survivor estimates. *Economica* 40, 85–94.

Rogers, R.J. (1993): The minimum optimal steel plant and the survivor technique of cost estimation. *Atlantic Economic Journal* 21, 30–37.

Salinger, M.A. (1984): Tobin's q, unionization and the concentration profits relationship. *Rand Journal of Economics* 15, 27–34.

Sengupta, J.K. and Fanchon, P. (1997): *Control Theory Methods in Economics*. Kluwer Academic Press, Dordrecht.

Sengupta, J.K. (1998): *New Growth Theory*. Edward Elgar, Cheltenham.

Sengupta, J.K. (2003): *New Efficiency Theory*. Springer, Berlin.

Solow, R.M. (1957): Technical change and the aggregate production function. *Review of Economics and Statistics* 39, 312–20.

Solow, R.M. (1997): *Learning from Learning by Doing*. Stanford University Press, Stanford.

Spulber, D.F. (1981): Capacity, output and sequential entry. *American Economic Review* 71, 503–14.

Stigler, G.J. (1958): The economies of scale. *Journal of Law and Economics* 1, 54–71.

Stigler, G.J. (1968): *The Organization of Industry*. Richard Irwin, Illinois.

Suzumura, K. (1995): *Competition, Commitment and Welfare*. Clarendon Press, Oxford.

Young, A. (1928): Increasing returns and economic progress. *Economic Journal* 38, 527–42.

3
Stochastic Selection and Evolution

Market selection process determines the survival of firms in an industry and their growth or decay. What are the factors which affect the selection process? Does size matter, that is, do small firms grow faster than large firms? Since investment decisions for output growth involve uncertainty and irreversibility, the resulting unit costs are only partially known. Future demand is also uncertain. Do the stochastic factors affect the selection process in a significant way? Does risk aversion play a significant role in firm's decisions to invest (disinvest) and expand (contract) its size?

The theory of stochastic selection emphasizes the hypothesis that stochastic forces are vital in the evolution process and it takes several forms. First of all, the decisions to invest for growth or expansion involve uncertainty associated with technology and future demand and the possibility of potential entry. When investments are irreversible, the probability of large sunk costs or lock-in costs arises if the demand fluctuations are expected to be large. Cost of switching from one technology to another may be high in uncertain environments.

Decisions to invest or expand capacity are also constrained by risk aversion costs, which may cause firms to hold the option to invest rather than exercise it. Thus an increase in uncertainty over future demand makes the firm hold less capacity.

Schumpeterian dynamics emphasize the innovation process in the selection process. Technological innovations produce both substitution-cum-diffusion and evolution and these effects are generally nonlinear over time involving multiple equilibria. This innovation stream has been frequently viewed as a stochastic process evolving over time. Assuming the set of technologies to be discrete and large, it may be represented by a set of integers $(N_1, N_2, \ldots, N_i, N_{i+1}, \ldots)$, where N_i may denote the number of production units or plants using the technology *i*. The transition of plants from one technology to another may be viewed as a birth and death process, that is, a Markov process, where birth may involve new plans entering the system and death implies that plants close down.

The birth and death process model may be viewed in several frameworks, for example, genetic evolution theory, economic diffusions between sets of innovations and finally the market selection mechanism where new entrants compete with the established firms. We would discuss these stochastic formulations with their economic and policy implications.

3.1 Fisherian replicator dynamics

Evolutionary economic theory has used one of the natural tools of genetic evolution theory to explain the variety of patterns in the evolution of species in a natural and regulated environment. This tool is replicator dynamics, which is used to describe the population-based nature of evolutionary theories of selection, where the frequency of a species (i.e., technologies, innovation processes or firms in economic context) grows differentially according to whether it has above or below average fitness. If fitness in genetic theory is replaced by economic efficiency or core competence in organization theory, the replicator dynamics in firm growth may be interpreted as a theory of distance from mean dynamics, that is, firms with above (below) the industry mean efficiency would grow (decline).

This has two major implications for neoclassical economic theory. One is that the neoclassical model of "representation agents" do not apply in this framework which emphasizes variety and diversity. Secondly, in this evolutionary framework agents or firms with a variety of behavioral and strategic characteristics coexist and define a joint population distribution characterized by its statistical moments such as the mean, variance and skewness. However, the statistical moments of the population distribution change over time as in a stochastic process via the environmental (market) selection process, where the rate of change of the mean may be a function of variance for example.

A simple formalization of replicator dynamics in genetic evolution theory is as follows:

$$\dot{x}_i = Ax_i(E_i - \overline{E}), \quad i = 1, 2, \ldots, n \tag{3.1}$$

where $\bar{E} = \Sigma x_i E_i$ = mean fitness. Here in genetic evolution theory x_i represents the proportion of species i in a population of n interacting species, E_i is its "reproductive fitness" and $\bar{E}$ is the weighted average fitness level in the total population. In economic theory x_i may represent any of the following: proportion of firms or plants using technology i, the market share of total industry output for firm i, or the entry of firm group i in the current period. E_i may represent average cost, efficiency or profit rate and $\bar{E}$ may denote the corresponding industry average. R.A. Fisher (1930) originally investigated the dynamic form (3.1) under the assumption that E_i's are constant, in which case the system monotonically converges to a steady state population consisting of the species with highest fitness.

Several implications of the replicator dynamics are important in the dynamic process of selection and evolution of industry. First of all, one of the important statistical properties of replicator dynamics is Fisher's fundamental theorem, which says that the rate of change of mean fitness is proportional to the variance (weighted by x_i) of fitness or efficiency characteristics in the population:

$$d\bar{E}_x/dt = \dot{\bar{E}}_x = -\alpha V_x(E_i), \quad \alpha > 0 \tag{3.2}$$

Here α is a positive constant measuring the speed of adjustment. In economic models Metcalfe (1994) and Mazzucato (2000) used replicator dynamics to analyze the dynamics of firm size and efficiency in evolution of industry following a Schumpeterian innovation framework.

Second, a broad implication of the Fisher principle embodied in (3.1) and (3.2) is that patterns of growth or decay in a species (technology or firm) depend on their differential behavior in terms of fitness (efficiency or profitability). Thus the patterns of evolution of the genetic system (industry or market system) are basically determined by the interaction of different species (firms or technologies) and their structural constraints. For economic systems the technology innovation process may be viewed as a stochastic process in terms of transition probabilities. Nelson and Winter (1982) suggest a formulation, where the transition probability $p_{ij}(t)$ from technology or state i to state j at time t depends on the "distance" between the two technologies or two states:

$$d(i,j) = w_l|\log a_l(i) - \log a_l(j)| + w_k|\log a_k(i) - \log a_k(j)| \tag{3.3}$$

where $a_l(i)$, $a_k(i)$ are the labor and capital coefficients of output for technology i and w_l, w_k are the nonnegative weights of labor and capital coefficients. The hypothesis is that the transition probability is concentrated on technologies close to the current one and decreases rapidly with the distance, for example,

$$p_{ij}(t) = \alpha(d^* - d(i,j)) \tag{3.4}$$

where d^* is the critical maximal technological or innovation distance which can still be crossed under reasonable market conditions.

Third, the role of adaptivity of the selection process to the environmental factors (H) has been considered in the generalization of the Fisher principle. If N is the size of a population, then the Fisher theorem implies

$$\mathrm{d}(\ln N)/\mathrm{d}t \leq f(H)$$

$$\text{and} \quad \mathrm{d}^2(\ln N)/\mathrm{d}t^2 = \frac{\partial f}{\partial H}\dot{H} - g(H,\dot{H}) + A(H)[f(H) - \mathrm{d}(\ln N)/\mathrm{d}t] \tag{3.5a}$$

$$= f(H, \mathrm{d}H/\mathrm{d}t, \mathrm{d}(\ln N)/\mathrm{d}t) \tag{3.5b}$$

Thus the imbalance in growth rate x of population in an adaptive environment may be written as

$$x = f(H) - \mathrm{d}(\ln N)/\mathrm{d}t > 0 \tag{3.6}$$

Then the system (3.5) of equations may be rewritten as a system of two first order equations

$$\mathrm{d}(\ln N)/\mathrm{d}t = f(H) - x \tag{3.7a}$$

$$\mathrm{d}x/\mathrm{d}t = g(H,\dot{H}) - A(H) \cdot x \tag{3.7b}$$

This shows that there exist three reasons for growth rate changes $\mathrm{d}(\dot{N}/N)/\mathrm{d}t$:

(a) A change in the balanced growth rate $(\partial f/\partial H)\,\dot{H}$, which explains the contribution of the environmental factors;
(b) A new imbalance in the growth rate generated by a changing environment denoted by $g(H,\dot{H})$. It follows from Fisher's theorem that $g(\cdot) \geq 0, g(H,0) = 0$;
(c) An adaptation process that brings the growth rate towards its equilibrium value. Here the adaptation process is assumed to be proportional to the imbalance in the growth rate. Here the term $A(H)$ depending on the environment may be called adaptability. In a fixed

environment it is negative. It tends to bring the growth rate towards its equilibrium value.

If the population size is the only environmental parameter or a suitable proxy, then (3.5) can be written more simply as

$$d^2(\ln N)/\mathrm{d}t^2 = (\partial f/\partial N)\dot{N} - g(N,\dot{N}) + A(N)\left[F(n) - \frac{\mathrm{d}\ln N}{\mathrm{d}t}\right] \tag{3.8}$$

Note that the traditional logistic-type growth model can be viewed as a special case of (3.8) when $A \to \infty$ and $\mathrm{d}(\ln N)/\mathrm{d}t = f(N)$. To see this one may linearize the equation (3.8) around an equilibrium value N^* which is the root of the equation $f(N) = 0$. Since $g(N,\dot{N})$ is always nonnegative, it will have a local extremum at the equilibrium point and therefore a zero term for the first approximation. Letting $z = \ln(N/\dot{N})$ we get

$$\ddot{z} + (k + A^*)\dot{z} + A^*kz = 0 \tag{3.9}$$

where $k = [\partial f/\partial(\ln N)]_{N=N^*} > 0$ measures the strength of density-dependence around the equilibriuim point and $A^* = A(N^*)$ is the adaptivity measured at an equilibrium point. Clearly the solution of this equation (3.9) is

$$z(t)[1 - k/A^*]^{-1}(z_0 + \dot{z}_0/A^*)\mathrm{e}^{-kt} + (kz_0/A^* + \dot{z}_0/A^*)\mathrm{e}^{-A^*t} \tag{3.10}$$

where dot denotes time derivative and z_0 is the initial value of z at $t = 0$. This sum of two exponentials in the solution path (3.10) shows two effects: the overshooting of the equilibrium level and the temporary decrease in size due to adaptation followed by an increase and an asymptotic equilibrium at $N = N^*$.

Finally, the Fisherian dynamics may be viewed in terms of an economic innovation process affecting the entry and exit behavior in a dynamic market. Thus if $N(t)$ is defined as the cumulative number of adopters of new innovations (e.g., new technology, new product or new marketing strategy) we may formulate innovations as a diffusion process:

$$\mathrm{d}N(t)/\mathrm{d}t = (p + qN(t)/m)(m - N(T)) \tag{3.11}$$

Here m is the ceiling of $N(t)$, p the coefficient of innovation and q the coefficient of imitation. Assuming $F(t) = N(t)/m$, the fraction of potential adopters who adopt the produce or technology by time t, one type of diffusion model can be formulated as

$$\mathrm{d}F(t)/\mathrm{d}t = [p + qF(t)](1 - F(t)) \tag{3.12}$$

On assuming $F(t=t_0=0)=0$, one obtains from (3.12) by simple integration the following distribution function for representing the time-dependent aspect of the diffusion process:

$$F(t) = \left[1 - e^{-(p+q)t}\right]\left[1 + (q/p)e^{-(p+q)t}\right]^{-1} \tag{3.13a}$$

$$\text{or} \quad N(t) = mF(t) \tag{3.13b}$$

For this growth curve the point of inflection which is the maximum penetration rate, that is, $(\partial F/\partial t)_{\max}$ occurs at t^*, where

$$t^* = -(p+q)^{-1}\ln(p/q)$$

$$F(t^*) = (1/2) - (p/2q) \tag{3.14}$$

Bass (1969) developed and applied this model in several industries. This type of model has several economic implications. First of all, the case $p=0$ considers the imitation effect, where firms tend to imitate the invention process of other firms and/or industries. In this case the model may be more generalized as

$$dF(t)/dt = q(1-\theta)^{-1}F^{\theta}(t)(1-F^{1-\theta}(t))$$

$$\text{that is,} \quad F(t) = \left[1 - (1 - F_0^{1-\theta})e^{-qt}\right]^{1/(1-\theta)} \tag{3.15}$$

where θ is a positive constant $0 \le \theta \le 1$ and $F_0 = F(t=0)$. This model (3.15) shows both symmetric and nonsymmetric diffusion patterns and hence is more generally applicable to various types of evolution of industries. Secondly, if $\theta=0$ and $F_0=0$, then $F(t)$ and hence $N(t)$ follows a logistic time-path that has been empirically observed for many technological innovations such as hybrid corn in agriculture or VHS technology in telecommunications.

Thus the replicator dynamics of the Fisherian theorem has many useful implications for economic selection processes in industry. The asymmetry in such processes is basically due to product differentiation and technological dispersal across firms and industries.

3.2 Schumpeterian innovation processes

Recent times have seen a spectacular growth in information technology (IT) and this challenge of new technology has spread far and wide in a global setting. Two aspects of this evolutionary dynamics have featured prominently in recent discussions. One is the Schumpeterian model of innovations and growth. This involves the process of impulse and diffusion in the propagation of new technology. Schumpeter (1961) considered six sets of innovations of which the introduction of a new method of production or technology and the opening of a new market through the introduction of new products are the most important. Secondly the dynamic role of R&D investment in the strategy followed by a successful entrant or innovator has been discussed by Aghion and Howitt (1992) and others. In this approach the successful innovator produces the newly invented good and protected by the patent right helps to drive out the previous incumbents by undercutting their price and this innovator enjoys monopoly rents until driven out by the next innovation. However, research and the associated knowledge capital have to face uncertainty of future developments which have intrinsic costs.

We discuss here two key elements of evolutionary growth emphasized in Schumpeterian dynamics: new innovations in technology and the process of creative destruction. The first, often referred to as the "technology push" hypothesis by Kamien and Schwartz (1982) and Scherer (1984) emphasizes the scale economies' effect of knowledge capital on modern output, whereas the second, termed as "demand pull" hypothesis puts more stress on innovation being spurred by enlarged market opportunities. Recently Metcalfe (1994) analyzed the technology diffusion process in terms of the innovative output following a logistic time-path, where capacity growth balances the increased demand through continual declines in costs and prices associated with the new technology.

For the new technology to displace the old, its production capacity must be optimally built up and the demand must be switched from the old to the new technology through the market selection process. Recently Aghion and Howitt (1992) developed a model of creative destruction in Schumpeterian dynamics, which postulates that a successful innovator drives out the previous incumbent by undercutting his process and creating a local monopoly through a patent until driven out by the next innovator. This innovation follows a stochastic process, where the evolutionary growth in the leading edge technology occurs through the competitive market selection process. The latter process has been discussed in a logistic framework by the model developed by Metcalfe.

Two stochastic aspects of the innovation process appear to play a key role in Schumpeterian dynamics. First is the existence of learning-by-doing scale economies by which the first innovating firm to race down the learning curve gains a cost advantage. If the cost advantage is sufficiently large, its possessor may select to set prices so low as to deter the entry of other competitors. Thus the innovating firms may have an incentive to lead by introducing new products which involve learning-by-doing. Secondly, the aggressive pricing of new products may stimulate demand, which permits more production and hence more learning. The parameters of the learning curve however involve two types of stochastic elements. One is the lowering of the unit production costs due to R&D activity and the other is the impact of the intra-industry spillover. Thus R&D investment not only generates new information for enhancing productivity but also increases the firm's ability to assimilate and exploit information flows as they develop.

Our attempt in this section is to develop stochastic models of Schumpeterian diffusion processes and their implications for stochastic instability in the input and output time series.

3.2.1 A diffusion model

We consider first the model of a Marshallian diffusion process due to Metcalfe (1994), which treats the diffusion of technology as the development of the new technology which is substituted for the old. Schumpeter's emphasis on the development of a new product or technology focuses on this aspect of diffusion as an evolutionary process. Let $x(t)$ be the supply of a new commodity at time t, for which the market demand is x_d. This diffusion model assumes that output growth $(\dot{x}/x)$ is proportional to the profitability of the new technology, subject to the constraint that the unit cost depends on the scale of production of the new technology:

$$\dot{x}/x \geq \dot{x}_d/x_d \tag{3.16}$$

$$\dot{x}_d/x_d = b(D(p) - x) \tag{3.17}$$

Here b is the constant adoption coefficient, dot is the time derivative and $D(p)$ is the long run demand curve for the new commodity. If the growth of capacity is in equilibrium with the growth in demand and price is proportional to marginal cost $p = k\,c(x)$, we obtain a balanced diffusion path as a logistic model:

$$\dot{x}/x = (\alpha - \beta x) \tag{3.18}$$

$$\text{where} \quad \alpha = b(d_0 - c_0 d_1 k), \quad \beta = 1 - c_1 d_1 k$$

$$c(x) = c_0 - c_1 x, \quad D(p) = d_0 - d_1 p$$

The model assumes a flexible competitive price process $p = p(t)$ which varies so as to equilibrate the capacity growth with the long-run growth of demand. Clearly there exist here three sources of equilibrium output growth. First is the diffusion parameter b. The higher the diffusion rate of new technology, the greater the output growth. Second, if demand rises over time and the innovator has a forward looking view of market growth, it stimulates capacity growth. The learning curve effect here implies declining unit costs through increase in cumulative output volume. Finally, the marginal cost tends to decline due to knowledge spillover across different firms and industries. Letting the parameters of the marginal cost function $c(x) = c(x, \gamma) = c_0(\gamma) - c_1(\gamma)x$ depend on the spillover factor γ one could capture the impact of inter-industry and international spillover of knowledge in new information technologies. Antonelli (1995) used this spillover diffusion process to empirically assess the notion of key technologies that provide positive externalities to the rest of the system.

If the logistic model (3.18) is viewed in deterministic terms with all parameters given, then the solution of the equilibrium time-path appears as follows

$$x(t) = A[1 + Be^{-\alpha t}]^{-1} \tag{3.19}$$

$$\text{where} \quad A = \alpha/\beta, B = (\alpha/\beta)\left(\frac{1}{x(0)} - \frac{\beta}{\alpha}\right)$$

Metcalfe (1988) has drawn two important implications from this solution path. One is that during the diffusion process, if one solves for the equilibrium prices, then it shows that the new technology/information flow is more profitable to adopt than the old and thus the clear incentive to switch to the new technology is built into it, although the lack of familiarity may hold back ready adoption. Second, the time-path of relative substitution of the new technology in place of the old that is displaced also follows the logistic curve. But since the parameters are not all deterministic in the long run, the stochastic variations imply a nonlogistic substitution envelope. The precise shape of this envelope depends on the nature and magnitude of the parameter variations and their temporal incidences.

Stochastic variations in the diffusion and technology adoption parameters are very basic to Schumpeterian growth theory which seeks to explain economic growth as arising from the flow of industrial innovations and knowledge. It specifically traces the specifics of this innovating knowledge, how it grows, is adopted and displaces the old. Schumpeter's idea of "creative destruction" emphasizes the stochastic nature of the innovation flow, how it develops out of the new initiative of the innovative entrepreneurs, how it generates rents and how it aids the process of dispersal across industries and international markets.

Recently Thompson (1996) developed a Schumpeterian model of endogenous growth which relates the market value of a firm to its current profits and to its R&D expenditures, where the firm's relative knowledge level follows a stochastic differential equation. If we interpret x_t in (1) as the firm's relative knowledge index, that is, $x_t = q_{it}/Q_t$ where q_{it} is the knowledge index and $Q_t = \int_0^t q_i \, di$ is the economy-wide knowledge index, then the expected growth of an innovative firm's relative knowledge level in (2) follows an expected profit maximizing behavior, since the growth of knowledge $(\dot{x}_t/x_t)$ is proportional to the expected profitability. Now consider the evolution process of relative knowledge (x_t) in terms of stochastic birth (i.e., fertility) and death (i.e., mortality) rates and let us consider the following transitions with the probability rates in infinitesimal time dt given as

Transition	Probability rate
$x \to x+1$	$g_1 x - h_1 x^2$
$x \to x-1$	$g_2 x + h_2 x^2$

Let ε_1, ε_2 be the error terms assumed to be Poisson variables with independent means $(g_1, x - h_1 x^2)\,dt$ and $(g_2 x + h_2 x^2)\,dt$ respectively, where

$$dx_t = (\alpha x_t - \beta x_t^2)\,dt + d\varepsilon_1 - d\varepsilon_2 \tag{3.20}$$

Then we obtain the variance of dε as

$$\begin{aligned}\text{Var}(d\varepsilon) &= \text{Var}(d\varepsilon_1 - d\varepsilon_2) = \text{Var}(d\varepsilon_1) + \text{Var}(d\varepsilon_2)\\ &= [(g_1 + g_2)x_t - (h_1 - h_2)x_t^2]\,dt\\ &= \alpha x_t - \beta x_t^2, \quad \text{where } \alpha = g_1 + g_2, \beta = h_1 - h_2.\end{aligned}$$

We may easily calculate the asymptotic stochastic mean of the Poisson process model (4) as follows. If μ is the true mean and $\tilde{x}(t) = x(t) - \mu$, then we obtain

$$\tilde{x}(t + \Delta t) - \tilde{x}(t) = (\alpha x_t - \beta x_t^2)dt + d\varepsilon_1 - d\varepsilon_2.$$

On taking expectations of both sides and rearranging terms one obtains

$$\mu = (1/\beta)[\alpha - \beta\sigma_x^2/\mu] \tag{3.21}$$

where σ_x^2 is the asymptotic variance of the stochastic process x_t. This equation (3.21) captures the steady state characteristics of the Poisson stochastic process underlying the deterministic logistic model (3.18). Two salient features of recent endogenous growth theory are borne out by this stochastic formulation. One is the implication that the mean (μ) and variance (σ_x^2) of the knowledge process is negatively correlated if $\mu > \alpha/(2\beta)$. If the output process y_t is proportional to x_t, this implies that the output fluctuations measured by variance (σ_y^2) would have a negative impact on the mean output (μ_y). Binder and Pesaran (1996) have interpreted the output variance as the cost of demand fluctuations and hence its negative impact on mean output as the optimal response of supply due to risk aversion.

A second salient feature of the stochastic logistic model (3.21) is that it generalizes the linear birth and death process model which follows from the Poisson process model above. Consider for instance the output process y_t as above and let the transition probability $p_y(t) = \text{Prob}[y_t = y]$ for $y = 0, 1, 2, \ldots$ satisfy the standard Markovian assumptions. Then the transition probability satisfies the so-called Chapman–Kolmogorov equation

$$\mathrm{d}p_y/\mathrm{d}t = \lambda_{y-1}p_{y-1} + \mu_{y+1}p_{y+1} - (\lambda_y + \mu_y)p_y$$

where the parameters λ_y, μ_y which depend on the level of output are called birth rate (fertility effect) and death rate (mortality effect) parameters, since the former leads to positive growth (e.g., the effect of experience and R&D) and the latter to decay (e.g., displacement of the old technology). Now assume that the birth rate parameter declines with increasing y_t and the death rate parameter remains proportional to y_t^2, that is, $\lambda_y = a_1 y_t(1 - y_t)$, $\mu_y = a_z y_t^2$. Then the mean $m_t = E(y_t)$ and variance $v_t = E(y_t - m_t)^2$ of the output process follow the nonlinear trajectory as follows:

$$\begin{aligned} \mathrm{d}m_t/\mathrm{d}t &= (a_1 + a_2)\left[\frac{a_1}{a_1 + a_2}m_t - m_t^2 - v_t\right] \\ &= a_1 m_t - a m_t^2 - a v_t, \quad a = a_1 + a_2 \end{aligned} \tag{3.22}$$

If we normalize as $a = a_1 + a_1 = 1.0$, then the deterministic and the stochastic logistic models appear as follows:

$$\begin{aligned} \mathrm{d}y_t/\mathrm{d}t &= a_1 y_t - y_t^2 \\ \mathrm{d}m_t/\mathrm{d}t &= a_1 m_t - m_t^2 - v_t \end{aligned} \tag{3.23}$$

Clearly in the steady state one obtains $\partial m/dv < 0$, whenever $m > (1/2)a_1$. Thus the cost of fluctuations would tend to lower the mean output, whenever m exceeds half the parameter a_1, which represents the positive diffusion coefficient associated with the birth rate λ_y. Moreover the presence of a positive variance function v_t in (3.23) implies that the deterministic trajectory in (3.23) would be shifted downward in the stochastic model. Due to this shift the steady state value m^0 of mean output would be less than that of the deterministic model (y^0).

This type of nonlinear stochastic growth model has been generalized by Sengupta (1998) in a two-sector framework, where the new innovating sector is more productive than the old. In this case the growth equations analogous to (3.23) could be derived for the two stochastic outputs Y_t and Z_t as

$$\begin{aligned} dm_{11}(t) &= \lambda_1 m_{11}(t) - \mu_{11} m_{12}(t) - \mu_{12}\bar{m}_{11}(t) \\ dm_{21}(t) &= \lambda_2 m_{21}(t) - \mu_{21}\bar{m}_{11}(t) - \mu_{22} m_{22}(t) \end{aligned} \tag{3.24}$$

where $m_{11}(t) = E\{Y_t\}$, $m_{12}(t) = E\{Y_t^2\}$, $m_{21}(t) = E\{Z_t\}$, $m_{22}(t) = E\{Z_t^2\}$ and $\bar{m}_{11}(t) = E\{Y_t Z_t\}$.

If the two outputs are deterministic, then the bivariate growth model reduces to

$$\begin{aligned} \dot{Y}_t &= \lambda_1 Y_t - \mu_{11} Y_t^2 - \mu_{12} Y_t Z_t \\ \dot{Z}_t &= \lambda_2 Z_t - \mu_{21} Y_t Z_t - \mu_{22} Z_t^2 \end{aligned} \tag{3.25}$$

where dot over a variable denotes its time derivative.

Note that the interaction terms denoted by $\bar{m}_{11}(t)$ and $(Y_t Z_t)$ play a very significant role here in determining the bivariate growth process. Let Y_t and Z_t be the respective outputs for the new and the old technology. Then the Schumpeterian model of creative destruction can be characterized by this model specified by (3.24) and (3.25). For instance if the interaction parameters μ_{12} and μ_{21} are zero and both μ_{11} and μ_{22} are negative, then both sectoral outputs tend to grow exponentially. But if $\mu_{11} < 0$ and $\mu_{22} > 0$ and $\mu_{12} = 0$ with $\mu_{21} > 0$ the innovating sector grows exponentially, whereas the old technology sector tends to decline over time.

3.2.2 Technical progress and evolutionary adaptation

The profit oriented aspect of innovation dynamics so far analyzed above assumes an automatic line between the generation of technological progress or change and the effective increase of total factor productivity. However the recent advances in the economics of innovation and new technology have shown that the introduction of technological change and its full adoption in the economic system is a lengthy process of evolutionary adaptation. Recently Antonelli (1995) has captured this adaptivity aspect in terms of a modified neoclassical production function

$$Y(t) = A(t)K^a(t)L^b(t)I_K^c(t) \tag{3.26}$$
$$A(t) = f(I_K(t))$$

where $Y(t)$ is output, $K(t)$ and $L(t)$ are the usual capital and labor inputs and $I_K(t)$ is the stock of information capital. Due to the dependence of $A(t)$ on the stock of information capital, significant amounts of externalities or spillover effects may be generated. Diffusion of technology embodied in information capital is assumed to follow a logistic process

$$i_K(t)/I_K(t) = b(h - I_K(t)) \tag{3.27}$$

where b is the positive rate of diffusion and h is the ceiling level of information capital. It is clear in this formulation that the general efficiency of each Cobb-Douglas production function (3.26) shifts towards the right, as the overall level of information capital increases. To empirically implement this specification Antonelli (1995) estimated the regression function of the average rate of growth of labor productivity on five sets of variables such as GDP per capita, average investment to GDP ratio, ratio of total US patents, diffusion of information and communication technologies (DICT), and a catching-up variable for 29 representative countries over the period 1980–88. His estimates found the DICT variable to be highly significant in a statistical sense (t-value 2.116) and his overall results confirm the finding that the diffusion rates of key technologies in the communications and related fields have generated significant externalities through knowledge spillover effects all through the economic system.

We now consider a two-stage model of the Schumpeterian innovation process, where growth in the leading edge occurs from successful innovations. We propose a two-stage model of Schumpeterian dynamics by combining the two key elements: new innovations in technology and the process of creative destruction.

The first stage formulates a stochastic flow model for the innovation input in the form of knowledge capital. The underlying stochastic process follows a birth and death process mechanism, where birth represents new ideas and death implies the relative obsolescence or destruction of the old. Then the second stage formulates a dynamic adjustment model for the innovator, who is assumed to minimize a loss function based on the discounted stream of deviations of innovative inputs from their desired or target levels. The target level may be determined either from the steady state version of stochastic birth and death process model or by minimizing a steady state cost function.

Whereas the first stage captures the role of uncertainty in the process of "technology push", the second stage emphasizes the disequilibrium framework of a "demand pull" behavior when the innovative producer has to adopt a dynamic view of the world economy when confronted with the openness in international trade and intensive market competition. The key issue here is: which of the two forces, past history or future expectations will play a more dynamic role in economic growth? In Schumpeterian dynamics of growth, the future expectations or the forward looking view ought to play a more dynamic role.

Consider the process of growth of innovation inputs x_t in the form of research or knowledge capital. As in the demographic theory of population growth the expected change in x_t during a discrete interval of time t to $t+1$ can be defined in terms of the deterministic model

$$E(x_{t+1}) = x_t + B_t - D_t \tag{3.28}$$

where B_t and D_t are expected births and deaths during the interval. Denoting expected birth rates (or entry rates) and death rates (or exit rates) by b_t and d_t and assuming a simple "birth" and "death" process model, the mean and variance of the innovation input may be easily written as follows:

$$\begin{aligned} E(x_{t+1}) &= (x_t e^{b_t - d_t}) = e^{r_t} x_t, \quad r_t = b_t - d_t \\ \mathrm{Var}(x_{t+1}) &= \frac{b_t + d_t}{b_t - d_t}[e^{r_t} - 1]e^{r_t} x_t, \quad \text{if } b_t \neq d_t \\ &= 2b_t x_t, \text{if } b_t = d_t \end{aligned} \tag{3.29}$$

Like demographic uncertainty the birth and death rates may be considered as shocks to the production process characterized by $y_t = f(x_t)$.

Clearly if $b_t > d_t$, then the expected innovation process is unbounded and hence there can be persistence in output growth over time as in new growth theory. In Schumpeter's thought the most favorable context for innovation is oligopolistic competition with large firms and the most important competitive pressure comes from technical change which erodes any monopoly position of a firm that does not remain in the forefront of technical change and research frontier.

Given the stochastic innovation process (3.28) we assume that the birth and death flows are endogenously determined as

$$B_t = b_0 + b_1 x_t + b_2 x_{t-1}; \qquad D_t = d_0 + d_1 x_t + d_2 x_{t-1} \tag{3.30}$$

This implies that births and deaths are proportional to the current and past innovation flows. On combining (3.28) and (3.30) one obtains the fundamental equation of the innovation process over time

$$x_t = -\beta_0 + \beta_1 x_{t+1} + \beta_2 x_{t+1} \tag{3.31}$$

$$\text{where} \quad \beta_0 = (1 + b_1 - d_1)^{-1}(b_0 - d_0), \quad \beta_1 = (1 + b_1 - d_1)^{-1}(d_2 - b_2)$$

$$\beta_2 = (1 + b_1 - d_1)^{-1}.$$

Two important implications follow from this equation. One is the case $\beta_1 < 0$ with $\beta_2 > 0$, when the past innovation serves to pull down the current flow, while the immediate future pulls it up. The net result is the positive growth in innovation, since entry dominates exit. This occurs when $b_2 > d_2$ with $b_1 \geq d_1$. The second case occurs when $b_2 < d_2$ with $b_1 \geq d_1$, implying $\beta_2 > \beta_1 > 0$. This implies that the future impact through demand pull is much stronger than the past trend in innovations. Clearly the impact of the past history and the future expectations may be more generalized in terms of more than one period, for example,

$$x_t = -\beta_0 + \sum_{i=1}^{m} \beta_{1i} x_{t-i} + \sum_{i=1}^{n} \beta_{2i} x_{t+i} \tag{3.32}$$

Finally, we note that there exists for both cases a positive level of steady state $\bar{x}$ if $\beta_1 + \beta_2 > 1.0$. Thus one can empirically test if the impact of future expectations is stronger than that of past history.

An interesting implication of the second order difference equation (3.31) is that it can be related to the dynamic adjustment behavior of a rational innovating firm. This behavior involves an optimizing decision by the producer, who finds that his current factor uses are not consistent with the long run equilibrium path (x_t^*, y_t^*) as implied by the stochastic model (3.31). These values x_t^*, y_t^* of inputs and output may also be interpreted as target levels implied by the current input prices and their expected changes in the future. Sengupta (1998) has recently applied a quadratic adjustment cost model to explore the optimal time-path of demand for capital and labor inputs in the growth process of Japan over the period 1965–90. By following this model we postulate that the innovating firm minimizes the expected present value of a quadratic loss function as follows

$$\text{Min } E_t L$$

$$\text{where} \quad L = \sum_{t=0}^{\infty} \rho^t \left[(\tilde{X}_t - \tilde{X}_t^*)' A (\tilde{X}_t - \tilde{X}_t^*) + (\tilde{X}_t - \tilde{X}_{t-1})' B (\tilde{X}_t - \tilde{X}_{t-1}) \right] \tag{3.33}$$

where $E_t(\cdot)$ is expectation as of time t, ρ an exogenous discount rate, prime is transpose, A and B are diagonal matrices with positive weights and $X_t, \tilde{X}_t^*$ are the vectors of input level and their targets. Here the first component of the loss function is disequilibrium cost due to deviations from either the desired level or steady state equilibrium and the second component characterizes the producer's aversion to input fluctuations under market uncertainty. Kennan (1979) and more recently Callen *et al.* (1990) have applied this formulation to derive optimal input demand equations which incorporates the producer's response to market fluctuations. On carrying out the minimization in (3.33) for the research input $\tilde{x}_{1t}$ say, one may easily derive the optimal adjustment behavior as

$$\tilde{x}_{1t} = b_0 + b_1 \tilde{x}_{1t-1} + b_2 \tilde{x}_{1t+1} \tag{3.34}$$

$$\text{where} \quad b_0 = (a_1 + a_2 + a_2\rho)^{-1}(a_1 \tilde{x}_{1t}^*)$$

$$b_1 = (a_1 + a_2 + a_2\rho)^{-1} a_2$$

$$b_2 = (a_1 + a_2 + a_2\rho)^{-1}(\rho a_2).$$

Similar equations for other inputs can be derived. On comparing (3.31) and (3.34) one could easily draw some interesting implications. First of all, one could estimate this model (3.34) and if it turns out that $\hat{b}_2 > \hat{b}_1 > 0$, or $\hat{b}_2 > 0$ with $\hat{b}_1 < 0$, then the future expectations play a more dominant role than the past history. This is the normal response in a Schumpeterian growth process involving innovating firms and for countries with rapid growth episodes. This may be empirically tested. Secondly, on using the nonexplosive characteristic root μ_1 of the second order difference equation (3.34), the optimal adjustment equation can be written as

$$\Delta\tilde{x}_{1t} = \tilde{x}_{1t} - \tilde{x}_{1t-1} = c_1 - g_1\tilde{x}_{1t-1} + h_1 d_{1t} \quad (3.35)$$

$$d_{1t} = (1 - \mu_1)\sum_{s=0}^{\infty} \mu_1^s \tilde{x}^*_{1,t+s}.$$

Finally, the gap between x_{1t} and $\tilde{x}_{1t}$, that is, $x_{1t} = \tilde{x}_{1t} + \varepsilon_{1t}$ may be evaluated over time to test if the planned inputs converge to the expected trend following from the stochastic process model. Indirectly it would provide an empirical test of the rational expectations hypothesis which postulates a perfect foresight condition in the sense that the ε_{1t} process is purely white noise.

In order to test the Schumpeterian hypothesis on growth, two countries: Japan (1965–90) and Korea (1971–94) are selected in their recent periods, when growth was moderate to rapid. For Japan the statistical data on the various inputs are obtained from the *Statistical Yearbook of Japan* and other official publications. The details of the data set are discussed in Sengupta (1998). For Korea (1971–94) the main sources of time series data are the *Korean Statistical Yearbook* and the official *Reports on Employment of Graduates* and *Science and Technology*. Some details of this data set are described in Sengupta (1998).

Tables 3.1 and 3.2 present the estimated results for Korea for the optimal input demand equations (3.34) for the three inputs: labor ($l_t = \ln L_t$), real capital stock ($k_t = \ln K_t$), and real GDP ($y_t = \ln Y_t$) and also knowledge capital ($h_t = \ln H_t$). Here knowledge capital is measured by the proxy variable: skilled labor with education level at the junior

Table 3.1 Estimates of optimal input demand and output for Korea (1971–94)

	Intercept	$t+1$	$t-1$	$\bar{R}^2$	DW
Labor (l_t)	−0.363	0.357*	0.351*	0.980	2.491
Real capital (k_t)	−2.489	0.564**	0.533**	0.975	1.637
Knowledge capital (h_t)	2.015	0.512**	0.498**	0.985	2.15
Output (y_t)	−0.023	0.573**	0.496**	0.647	2.239
Export share (e_t) of GDP (in logs)	−0.882	0.593*	0.507	0.589	2.381

Notes
1. One and two asterisks denote significant t-values at 5 and 1% of two-sided tests and $\bar{R}^2$ is squared multiple correlation coefficient adjusted for degrees of freedom.
2. The estimates are corrected by the Cochrane–Orcutt procedure and the independent variables $x_i(t+1)$ are estimated by Kennan's procedure.

Table 3.2 Growth of R&D and its impact on growth of real GDP in Korea (1971–94)

$\Delta R_t/R_{t-1} = b_0 + b_1 R_{t-1}$					
	b_0		b_1	ρ	$\bar{R}^2$
R&D expenditure					
Level	0.156*		−0.002	0.341	0.094
Per worker	0.231		−0.014	0.320	0.089
$\Delta y_t/y_{t-1} = a_0 + a_1 \Delta k_t/k_{t-1} + a_2 R_t/R_{t-1}$					
R&D expenditure					
Level	0.007	0.494**	−0.026	−0.159	0.345
Per worker	0.0008	0.531**	−0.005	−0.241	0.381

Notes

1. Total R&D expenditure (R_t) is used as a proxy for knowledge capital.
2. One and two asterisks denote as before (Table 3.1).
3. All the estimates are corrected by the Cochrane–Orcutt procedure.

Table 3.3 Estimates of input and output demand equations for Japan (1965–90)

	Intercept	$t+1$	$t-1$		$\bar{R}^2$	DW
Skilled labor (s_t)	1.045	0.589**	0.402**		0.912	2.712
Unskilled labor (u_t)	1.124	0.471**	0.465**		0.813	2.581
Capital (k_t)	−0.026	0.512**	0.489**		0.893	2.675
Output (y_t)	0.039	0.495**	0.501**		0.735	2.728
$\Delta y_t = a_0 + a_1 \Delta k_t + a_2 \Delta s_t + a_3 \Delta u_t$						
	a_0	a_1	a_2	a_3	$\bar{R}^2$	DW
Δy_t	−0.003	0.545*	0.127	−0.162	0.728	1.748

Notes

1. One and two asterisks denote as before (Table 3.1).
2. All the estimates are corrected for heteros cedasticity by the Cochrane–Orentt procedure.

college level or higher. Table 3.3 presents the estimated results for the optimal input demand equations for Japan, where the role of knowledge capital is captured by the skilled ($s_t = \ln L_t^s$) and unskilled labor force ($u_t = \ln L_t^u$), skill being defined by the junior college level education at the minimum or higher.

Two important implications follow from the estimates of Tables 3.1 and 3.2 for Korea. First of all, the forward looking view of the future expectations represented by $x_{i,\,t+1}$ is more dominant than the past represented by $x_{i,\,t-1}$. This is particularly so for the export share (e_t), real capital and knowledge capital. For the period 1971–94 covered here this tendency has shown more persistence. For example when we run regressions of the type (3.32) with higher order variables such as $x_{t,\ t+2}$ an $x_{i,\ t-2}$ the coefficients turn out to be insignificant, but the dominance result still persists. Furthermore, if we estimate the demand for knowledge capital in the form (8) it turns out to be

$$\Delta h_t = 1.072 - 0.512\, h_{t-1} + 0.019t \quad \overline{R}^2 = 0.987$$

showing the trend is one of decline. When a logistic model is fitted to the h_t time series, it shows the rate of decline more prominently. Second, if we measure research innovations by R&D expenditure as in Table 3.2, the long-run decline is evident from the coefficient of R_{t-1} on the growth of R&D expenditure. Thus the forward looking view on growth has not been sustained by any increase in long-run trend of growth in knowledge capital either in the form of educational skills or R&D expenditure. This tends to support Young's (1995) view that the East Asian growth experience has not been self-sustained through the growth in knowledge capital. Finally, we have the optimal input demand estimates for Japan (1965–90) in Table 3.3, which shows the dominance of the forward looking view over the past though in a much slower fashion. When higher order terms are allowed in the regression model the results appear as follows:

$$\Delta k_t = -0.731 - 0.282^{*}\, k_{t-1} + 0.787^{**}\, k_{t+1} - 2.420^{**}\, k_{t-2} - 0.257^{*}\, k_{t+2}$$

$$\overline{R}^2 = 0.924$$

$$\text{DW} = 2.397$$

$$\Delta y_t = 0.038 - 0.387^{*}\, y_{t-1} + 0.743^{**}\, y_{t+1} - 0.155\, y_{t-2} - 0.205\, y_{t+2}$$

$$\overline{R}^2 = 0.719$$

$$\text{DW} = 2.893$$

The table also shows that skilled labor contributes positively to output growth, though this is not found to be statistically significant.

3.3 Stochastic models of entry

The entry and exit dynamics of firms are basically affected by the stream of innovations and their effects on price and costs. Innovations here have to be interpreted in a very broad sense including the following: (a) production efficiency and its improvement by restarting the cycle in price and quality area, (b) technological efficiency through timing and know-how improvement through R&D investments, (c) access efficiency in the strongholds arena and networking and (d) improving resource and product efficiency. The stream of innovations involves uncertainty in future developments and switch-in costs. Also sunk costs are important in inhibiting the displacement process by which less efficient plants are replaced by more efficient ones.

The concept of entry can be defined and measured in a number of ways. Three types of measures would be used here in our inquiry into the stochastic aspects of entry. The first measure defines entry as the increase in output of the new entrants over time who compete with established firms in the industry. Taking these two groups of firms, the new entrants and the incumbents, one could visualize the market dynamics as a two-player differential game. Gaskins (1971) and others have formulated limit pricing models in this framework. A second measure of entry is defined by the net change in the number of firms in an industry. Thus Dreze and Sheshinski (1984) have considered dynamic adjustment models which assume that new plants of a given type i are built whenever their expected profits are positive but that plans of type i are scrapped or at least not replaced when their expected profits are negative. The third definition considers entry in terms of the relative absence of barriers to entry (BTE). The proportion of new entrants is here related to BTE. The BTE is a vector which includes product differentiation (i.e., patents), economies of scale (minimum efficient scale) and the degree of sunkness of machinery and equipment capital. Thus Mata (1995) has constructed econometric models of entry and exit and also a model of new entry, that is, entry over exit where the BTE vector is used as explanatory variables.

We consider here these three types of entry models and discuss the role of various stochastic factors. These models have interesting dynamic properties of stability and various econometric applications of these models have been made for several types of industries like computers, microelectronics and manufacturing.

We consider first the limit pricing model, where the limit price p_0 is defined as the highest common price which the established firms believe they can charge without inducing any new entry or market penetration. In other works $p_0 > 0$ is the entry preventing price. In a competitive market this may be very close to the minimum point of the long run average cost. Market entry by rivals is measured by $\dot{x} = dx/dt$, where

$x = x(t)$ is their supply. The entry is positive (negative) if the market price $p(t)$ exceeds (falls below) p_0. The entry dynamics is then represented as

$$\dot{x} = h(p(t) - p_0), \ p(t) \geq p_0 \tag{3.36}$$

where $h(\cdot)$ is a continuous nondecreasing function such that there is no entry ($\dot{x} = 0$) when $h = h(0)$, $p(t) = p_0$. The market clearing price follows the demand function

$$p(t) = \mathrm{d}(x(t), q(t)) \tag{3.37}$$

where $\mathrm{d}(\cdot)$ in general is a nonlinear function and $q(t)$ is the quantity supplied by the established firms, which are viewed as a dominant firm in the Gaskins model. The dominance is measured by the large market share of total industry supply given by $x(t) + q(t)$. On combining (3.36) and (3.37) we obtain the dynamic entry equation:

$$\dot{x}(t) = f(x(t), q(t); p_0) \tag{3.38}$$

where $f(\cdot)$ is a nonlinear function of its arguments. Consider the simple case when $q(t)$ is set by the dominant firm group as a leader, who knows that the new entrants would follow their reaction curves: $R_x : x(t) = R(q(t))$. Then one can write the entry equation as

$$\dot{x}(t) = f(x(t), t)$$

by incorporating the reaction function R_x and subsuming the constant limit price p_0. But since entry is uncertain one has to introduce a stochastic generating mechanism. One useful approximation is the linearized form:

$$\dot{x}(t) = -(\alpha \mathrm{d}t + \mathrm{d}B(t))\, x(t) \tag{3.39a}$$

where α is assumed to be a positive constant reflecting market growth and the slope of the demand function in the steady state and B(t) is assumed to be a stationary Brownian motion with mean zero and variance ($\sigma^2 t$). It is true that the assumption on the stochastic process $B(t)$ here is somewhat strong, since in the real world nonstationarity may be present. But even with such assumptions the problems of instability of the entry dynamics cannot be ignored. The stochastic differential equation (3.39a) has the solution

$$x(t) = x(0) \ \exp[-(\alpha + \sigma^2/2)t - B(t)] \tag{3.39b}$$

Since the Brownian process $B(t)$ grows like $(t \log \log t)^{1/2}$ with probability one, the stability property of the solution path (3.39b) is determined by the deterministic term in the exponent of (3.39b), that is, $a > 0$. But the region of sample stability is specified by

$$\hat{\alpha} + \hat{\sigma}^2/2 > 0 \tag{3.39c}$$

where a hat over the variable denotes its sample estimate. The variance of the $x(t)$ process can be explicitly computed here for the linearized process (3.39a) as:

$$\text{Var}(x(t)) = x^2(0)\{\exp[-(2\hat{\alpha} - \hat{\sigma}^2)t] - \exp(-2\hat{\alpha}t)\}$$

Clearly we need the condition $2\hat{\alpha} > \hat{\sigma}^2$ for the variance to be stable as $t \to \infty$, since we have

$$E|x(t)|^k = \exp[(1/2)\, kt(k-1)\, \hat{\sigma}^2 - 2\hat{\alpha}].$$

Therefore for the stability of $E|x(t)|^3$ and $E|x(t)|^4$ we need the respective conditions of

$$\hat{\alpha} > \hat{\sigma}^2 \text{ and } 2\hat{\alpha}/3 > \hat{\sigma}^2.$$

It is apparent that as k increases ($k = 1, 2, 3, 4, \ldots$), the stability region decreases and hence the satisfaction of the condition (3.39c) for stability of the mean entry process does not guarantee that the variance and higher moments would be stable in the sense of being bounded as $t \to \infty$.

Two broad implications of stochastic instability of the linearized entry process may be noted. One is that the high variance of potential entry (or exit) may imply a large cost of dynamic risk aversion in the form of sunk cost and lock-in cost. Secondly, the linearized generating mechanism (3.39a) can be further improved in its economic content, if we view the entry probability as an increasing convex function of price at time t, given that entry has not yet occurred. This has been the approach of Kamien and Schwartz (1971). Let $F(t)$ denote the probability that entry has occurred by time t with $F(0) = 0$. Then the conditional probability of entry at time is $\phi(t) = \dot{F}(t)/(1 - F(t))$, where $\dot{F}(t) = dF/dt$. Hence the modified entry equation becomes

$$\phi(t) = h(\tilde{p}(t)), \quad \tilde{p}(t) = p(t) - p_0 \tag{3.40}$$

with $h(0) = 0$, $\partial h/\partial p \geq 0$ and $\partial^2 h/\partial p^2 \geq 0$. The economic interpretation is that a higher expected price $\tilde{p}(t)$ increases the probability of entry (or, shortens the expected entry lag). Note that if the $h(\cdot)$ function is decomposed into new innovation and imitation components as $h(\tilde{p}(t)) = (\phi_1 + \phi_2 F(t))$, that is, ϕ_1 is the coefficient of innovation which affects average costs and price and ϕ_2 is the coefficient of imitation which saves average research cost per unit of $F(t)$, then this entry process (3.40) yields the Bass-type diffusion model (3.12) we considered before, that is, $\mathrm{d}F(t)/\mathrm{d}t = [\phi_1 + \phi_2 F(t)](1 - F(t))$ which yields the solution path

$$F(t) = [1 - \exp(-t(\phi_1 + \phi_2))][1 + (\phi_2/\phi_1)\exp(-t(\phi_1 + \phi_2)]$$

Now consider a more general formulation of the entry process. Let y_i be the mean market share of firm i ($y_i = Ex_i$), where market penetration is viewed as potential entry. Then the differential equation (3.38) becomes

$$\dot{y}_i(t) = f(y) \tag{3.41}$$

where $f(y)$ is an n-vector valued continuous function defined on a set W of n vectors $y = (y_1, y_2, \ldots, y_n)$. We make the assumption (A) that for any initial position y^0 in W the entry process (3.41) has a solution $y(t; y^0)$ for all $t \geq 0$ which is uniquely determined by y^0. Let $\bar{y} = (\bar{y}_1, \ldots, \bar{y}_n)$ in W be called an equilibrium point if $f(\bar{y}) = 0$. Clearly the set E of all such equilibrium points is closed in W.

The entry process (3.41) is defined to be (globally) *stable*, if for any initial position y^0 in W, the solution $y(t; y^0)$ to the entry process converges to an equilibrium $\bar{y}$. But the entry process (3.41) is called *quasi-stable*, if for any initial position y^0 in W, the solution $y(t; y^0)$ is bounded and every limit point of $y(t; y^0)$ as $t \to \infty$ is an equilibrium.

One can now prove the following stability theorem:

Theorem of Stability
Let $f(y)$ in the entry process (3.41) satisfy assumption (A) and the following

(B) For any initial position y^0 in W the solution $y(t; y^0)$ to the entry process (3.41) is contained in a compact set W and
(C) there exists a continuous function $\psi(x)$ defined on W which is a strictly decreasing function with respect to t unless $y(t; y^0)$ is an equilibrium point.

Then the dynamic entry process (3.41) is *quasi-stable*.

Proof: Let $u(t) = \psi[y(t; y^0)]$

By assumption (B) the set $\{y(t) : 0 \le t < \infty)$ is contained in a compact set in W. By assumption (C), $u(t)$ is a nonincreasing function of t. Hence $\lim_{t\to\infty} u(t)$ exists and is equal to say u^*. By assumption (B) a limit point $y^* = \lim_{\tau\to\infty} y(t_\tau)$ is contained in the set W. Now consider the solution $y^*(t) = y(t; y^*)$ to the entry process (3.41) with initial position y^* and let $u^*(t) = \psi[y^*(t)]$. Since $\psi(y)$ and $y(t; y^0)$ is continuous with respect to y and y^0 respectively, we get

$$y^*(t; y(t_\tau)) = y(t + t_\tau; y^0).$$

$$\text{Hence} \quad u^*(t) = \lim_{\tau\to\infty} \psi[y(t + t_\tau); y^0]$$

$$= \lim_{\tau\to\infty} u(t + t_\tau) = u^*, \quad \text{for all } t \ge 0$$

Therefore by condition (C) we obtain $y^* = y^*(0)$ as an equilibrium point. This proves that the entry process (3.41) is quasi-stable.

Note that the function $\psi(y)$ satisfying assumption (C) may be viewed as Lyapunov distance function, for example,

$$\psi(y) = \min_{\bar{y}\in E} |y - y^*|^2$$

where E is a closed set defined before.

Two implications of this stability theorem are important. One is that the distance function implies by assumption (C) that

$$\sum_{i=1}^{n} (\partial\psi/\partial y_i) f_i(y) < 0$$

for all nonequilibrium points y in the set W. Second, the quasi-stable region of the mean entry process may be amenable to empirical estimation from the observed data set. However, one has to note that there may exist other regions of instability in the stochastic case, when the variance process associated with the entry dynamics may not be stable in the sense discussed before.

We now consider a second view of stochastic entry, which is more closely related to the stochastic specification of the market demand function. In the demand function (3.37) the price variable is assumed to be stochastic and hence the probability $\text{Prob}(p(t) > p_0)$ has an impact on the entry equation (3.36). Assume a linearized model with a demand function

$$p(t) = a - b(q(t) + x(t)) + e(t)$$

where $e(t)$ is the noise component following a specified zero mean stochastic process. By using the probabilistic inequality $\text{Prob}(p(t) \geq p_0) = \varepsilon$ and the cumulative distribution $F(s)$ we may replace $p(t)$ by $Ep(t) - \alpha\sigma(t)$ where $Ep(t)$ and $\sigma^2(t)$ are the mean and variance of price and $\alpha = -F^{-1}(1-\varepsilon)$, if the distribution is normal and $\varepsilon > 0.50$. The entry equation then becomes

$$\dot{x}(t) = h(Ep(t) - \alpha\sigma(t) - p_0).$$

If the potential entrants are risk averse (higher positive value of α), then positive entry occurs only when

$$Ep(t) > p_0 + \alpha\sigma(t)$$

Thus the potential barrier to entry enters into the entry dynamics through higher costs of risk and the latter may be due to higher probability of sunk costs and lock-in or switching costs.

3.3.1 Cost adjustment models of entry

Cost based models of entry emphasize the stochastic nature of costs due to uncertainty associated with R&D investment, technological innovations and market uncertainty. If the cost frontier includes all static and dynamic costs, then those firms which fail to remain on the dynamic cost frontier may feel the pressure of competition. This increases their probability of exit. If entry is viewed as market penetration, then the new entrants may increase their long-run market share if they can reduce their long-run average costs below the industry average.

Consider an industry comprising numerous firms producing a single good. Firms are identified by their cost structure, where each firm follows one of K possible types of cost frontiers. This type of diversified cost structure has been used by Dreze and Sheshinski (1984). Let n_i be the number of firms with costs $C_i(y_i), i = 1, 2, \ldots, K$ where each cost frontier is assumed to be monotone increasing, strictly convex and quadratic for simplicity

$$C_i = C_i(y_i) = \gamma_0 + \gamma_1 y_i + \gamma_2 y_i^2 + e_i \tag{3.42}$$

Here $\gamma_0, \gamma_1, \gamma_2$ are positive parameters and e_i is a stochastic term distributed independently with zero mean and variance σ_i^2. The total costs for the industry are given by

$$C = \sum_{i=1}^{K} n_i C_i(y_i) \tag{3.43a}$$

The industry selects the optimal number of firms n_i^* by minimizing the expected value of C in (3.43a) subject to

$$\sum_{i=1}^{K} n_i y_i \geq D; \quad n_i \geq 0 \tag{3.43b}$$

where D is market demand assumed to be given y consumer preferences. On using p as the Lagrange multiplier for (3.43b), the optimal solution (n_i^*, p^*) must satisfy the Kuhn–Tucker necessary conditions as

$$EC_i(y_i) - p(n, D) \geq 0; \qquad y_i[C_i(y_i) - p(n, D)] = 0, \quad \text{all } i \tag{3.43c}$$

$$\text{and} \quad \sum_{i=1}^{K} n_i y_i = D, \quad \text{all } i = 1, 2, \ldots, K$$

Clearly $p^* = \overline{C}_i^* / y_i = E(AC_i)$ and since each C_i is a cost frontier, hence $p^* = E(MC_i)$, where E is expectation and AC, MC denote average and marginal cost. Hence optimal price p^* equals the minimum of $E(AC_i)$. Thus the long-run equilibrium price equals the average cost at the minimum efficient scale (MES) of output. Two implications of this result are important. One is that the entry or market penetration would increase for any firm which can successfully reduce the MES level of average cost and hence its equilibrium price. Secondly, if market demand rises so as to increase the market price above the MES level of average cost, then the probability of new entry would rise. This is because the expected profits $E\pi_i$ defined by

$$E\pi_i = E[y_i p(n, D) - C_i(y_i)]$$

would rise. Thus the entry dynamics follows the process:

$$dn_i/dt = \begin{cases} k_i E\pi_i, & k_i > 0, \ n_i > 0 \\ \max\{0, k_i E\pi_i\}, & n_i = 0 \end{cases} \tag{3.44}$$

This is a linearized version, although it would hold for any strictly increasing continuous function $h_i(E\pi_i)$ with $h_i(0) = 0$.

This type of entry assumes that new plants of type i are built whenever their expected profits are positive and vice versa. Since expected profits in the entry equation (3.44) can be expressed in terms of the vector $n = (n_i, n_2, \ldots, n_K)$ the differential process can be written as

$$\dot{n}_i = f(n)$$

Then the theorem of stability developed in (3.41) can be applied here, implying the existence of the stability zone under the standard demand functions with a negative slope and strictly convex cost functions. We have to note also that n_i is assumed here to be continuous but this is an approximation and for convenience.

The above formulation of the stochastic entry process may be generalized in two ways. One is to assume that each firm is risk averse and minimizes the adjusted expected cost function

$$EC_i = \overline{C}_i + \alpha\sigma_i^2, \quad \alpha > 0$$

where α is the weight on variance of costs. In this case the optimal price p^* becomes

$$p^* = (\overline{C}_i + \alpha\sigma_i^2)/y_i > \overline{C}_i/y_i.$$

Thus unless demand drives the market price above the level p^*, there is no incentive for entry. Second, one may assume market demand D to be stochastic instead of costs $C_i(y_i)$. A similar entry equation of the form (3.44) may be easily derived.

3.3.2 Barriers to entry

The econometric model developed by Mata (1995) and others considered entry in terms of barriers to entry (BTE). If BTE exists to a significant degree, then it prevents potential entry. They consider BTE in several forms: existence of patents, large scale of plant (MS), large sunk cost and product differentiation and find empirically a statistically significant role of the scale of plant or MSE. If entry is viewed as market penetration, then one formulates the entry process in terms of the expected entry variable E_i for firm i in terms of its market share ($i = 1, 2, \ldots, n$) as follows

$$\begin{aligned} &\mathrm{d}E_i/\mathrm{d}t = \lambda(I_i)[\bar{c} - c_i] \\ &\lambda = \lambda(I_i), \quad I_i = \text{innovations} \\ &\bar{c} = \text{industry average cost} \end{aligned} \tag{3.45}$$

c_i is the minimal average cost of firm group i representing BTE. Thus if BTE rises, it tends to restrict entry. One may also view c_i as a function of innovations, so that the efficient firms may innovate to a significant degree which may continually reduce c_i over time. This restricts the long-run prospect of entry. In models of hypercompetition this has been characterized as resource efficiency. Note that the parameter λ here determines the speed at which firms' market penetration reacts to differences between firm efficiency characteristics. As an industry-specific parameter, a high value of λ would describe an industry with a strong competitive adjustment mechanism. This may intensify the impact of BTE on entry prevention.

Thus the stochastic basis of the entry and exit dynamics affects to a significant degree the pattern of evolution of an industry. This pattern is far from the proportional growth law (e.g., Gibrat's law) since there are asymmetric characteristics in the uncertain environment involving the mean, variance and skewness parameters.

References

Aghion, P. and Howitt, P. (1992): A model of growth through creative destruction. *Econometrica* 60, 323–51.

Antonelli, C. (1995): The diffusion of new information technologies and productivity growth. *Journal of Evolutionary Economics* 5, 1–17.

Bass, F.M. (1969): A new product growth model for consumer durables. *Management Science* 15, 215–27.

Binder, M. and Pesaran, M.H. (1996): Stochastic growth. Working paper, Department of Economics, University of Maryland, Maryland.

Callen, T., Hall, R. and Henry, S. (1990): Manufacturing stocks, expectations, risk and co-integration. *Economic Journal* 100, 756–72.

Dreze, J. and Sheshinski, E. (1984): On industry equilibrium under uncertainty. *Journal of Economic Theory* 33, 88–97.

Fisher, R.A. (1930): *The Genetic Theory of Natural Selection*. Clarendon Press, Oxford.

Gaskins, D.W. (1971): Dynamic limit pricing: Optimal pricing under threat of entry. *Journal of Economic Theory* 3, 306–22.

Kamien, M.I. and Schwartz, N.L. (1982): *Market Structure and Innovation*. Cambridge University Press, Cambridge.

Kennan, J. (1979): The estimation of partial adjustment models with rational expectations. *Econometrica* 47, 1441–57.

Mata, J. (1995): Sunk costs and the dynamics of entry in portuguese manufacturing. In: Witteloostuijn, A.V. (ed.), *Market Evolution*. Kluwer Academic, Dordrecht.

Mazzucato, M. (2000): *Firm Size, Innovation and Market Structure*. Edward Elgar, Cheltenham.

Metcalfe, J.S. (1994): Competition, evolution and the capital market. *Metroeconomica* 45, 127–54.

Nelson, R. and Winter, S. (1982): *An Evolutionary Theory of Economic Change*. Harvard University Press, Cambridge.

Scherer, F.M. (1984): *International High-technology Competition*. Harvard University Press, Cambridge, MA.

Schumpeter, J.A. (1961): *The Theory of Economic Development*. Oxford University Press, New York.

Sengupta, J.K. (1998): *Growth Theory: An Applied Perspective*. Edward Elgar, Cheltenham.

Thompson, P. (1996): Technological opportunity and the growth of knowledge: a Schumpeterian approach to measurement. *Journal of Evolutionary Economics* 6, 77–97.

Young, A. (1995): The tyranny of numbers: confronting the statistic realities of the East Asian growth experience. *Quarterly Journal of Economics* 110, 641–80.

4
Cournot–Nash Selection Process

The selection process of firms in the evolution of industry follows a Cournot–Nash equilibrium, when firms operate in an oligopolistic market. We consider first a static symmetric framework where each of n firms operates under constant cost c and market demand is described by a stationary inverse demand curve, which is assumed to be linear for simplicity. This static model is then generalized to a noncooperative dynamic game where the firms are viewed as Cournot players, each maximizing the present discounted value of its profit stream over the horizon $[0, \infty]$ at the constant positive rate of discount r. This framework naturally yields a differential game formulation, where the role of information available to each player plays a significant role. Asymmetry of the information structure can also be introduced at this stage, along with the stochastic aspects in the cost and demand structure.

4.1 Static framework and entry

Consider a symmetric Cournot framework with identical unit cost c, where the unique Cournot–Nash equilibrium is defined for each firm as $y_j(c)$, where $p = p(D)$, $D = ny_j$ is the market clearing price with market demand as D.

$$p(ny_j(c)) + y_j(c)p'(ny_j(c)) - c = 0 \tag{4.1}$$

where prime denotes the slope of the inverse demand curve. It is easy to show that $y_j(c)$ is decreasing in n and c and that $ny_j(c)$ is increasing in n and decreasing in c. With a linear demand function $p = a - b(ny_j(c))$, one obtains

$$y_j = (a - c)[b(n + \beta_j)]^{-1}, \quad \beta_j = \partial Y/\partial y_j, \quad Y = \sum_{j=1}^{n} y_j$$
$$= (a - c)[b(n + 1)]^{-1}, \quad \text{if } \beta_j = 1, \text{ all } j \tag{4.2}$$

This formulation ignores the role of R&D and its impact on innovative activity for each firm. To make R&D expenditure as an explicit part of a firm's strategy we may follow the model due to Dasgupta and Stiglitz (1980). Here the profit of each firm is written as

$$\pi_j = p(D)y_j - c(x_j)y_j - x_j \tag{4.3a}$$

where x_j is the R&D spending which benefits the firm in lowering its unit costs $c(x_j)$. The optimal amount x_j^* of R&D expenditure must satisfy the optimality conditions:

$$\partial\pi_j/\partial x_j = 0, \quad -[\partial c(x_j^*)/\partial x_j]y_j^* = 1 \tag{4.3b}$$
$$\partial\pi_j/\partial y_j = 0 \text{ yielding } [p - c(x^*)]/p = s_j/e.$$

Here s_j is the share of firm j in total industry output and e is the elasticity of total market demand. Some economic implications of the conditions (4.3b) may be noted. First of all, the first equation of (4.3b) implies that the marginal benefit of an extra dollar of R&D spending must equal in equilibrium to the marginal cost which is simply $1. Thus if $-\partial c(x_j^*)/\partial x_j) > 1$ then firm j should spend more on R&D. The second part of equation (4.3b) says that if the elasticity of market demand declines as output expands as is the case with linear demand curves, then increasing the number of firms will, beyond some point lead to a reduction in total R&D spending. Second, if free entry conditions prevail in the long-run equilibrium, then free entry leads to an increase in the number of firms until each firm makes zero profits. Applying this zero profit condition to (4.3a) and summing over n firms we obtain

$$[p(Y^*) - c(x^*)]Y^* = n^*x^* \tag{4.3c}$$

where $Y = ny_j = \Sigma y_j$. Hence the result

$$(n^*x^*)/(p(Y^*)Y^*) = (n^*e)^{-1}$$

which says that the industry R&D spending as a share of industry sales is inversely proportional to the optimal number of firms n^* and the market price elasticity of demand. Thus the industries with a naturally more competitive (less concentrated) structure will do less R&D, other things being equal. This provides some theoretical support for Schumpeter's hypothesis that imperfect competition or oligopoly is good for technical progress and innovations. However, the empirical evidence on this hypothesis does not provide a clear cut answer, implying the existence of other significant factors besides the degree of monopoly power, for example, the nature of technical innovation.

The zero profit condition (4.3c) has two implications for entry. One is that the existence of positive profits would increase the probability of the entry. Thus the limit pricing model framework holds here, for example.

$$dn/dt = \dot{n} = k(\pi - \pi^*), \quad \pi^* = 0, \; k > 0 \tag{4.4}$$

Second, the degree of market power held by a firm (i.e., a dominant firm) may delay the period of the long-run framework, just as other barriers to entry (BTE) may help to postpone potential entry. The entry dynamics here follow the competitive process of Walrasian adjustment, where $\pi > \pi^* = 0$ invites entry, $\pi < 0$ invites exit and $\pi = 0$ implies no entry ($\dot{n} = 0$).

Novshek (1980) explored this type of Cournot–Nash equilibrium under conditions of free entry and showed the validity of the *folk theorem* which says that if the oligopolistic firms are small relative to the total market, then the market outcome is approximately competitive. Hence the competitive process entry model of Walrasian adjustment (4.4) holds approximately. This result is important for two reasons. One is that the Novshek formulation relates the absence of entry to the existence of minimum average cost attained by the incumbent firms. Second, it showed that if firms are significant but small, they can be assumed to recognize their market power on price and in this framework a Cournot–Nash equilibrium with free entry still exists. Then the resulting oligopolistic market outcome is approximately perfectly competitive. This provides a justification for using the long-run perfectly competitive model with infinitesimal firms as an idealization of markets with free entry where firms are technologically small relative to the total market.

When the conditions of the folk theorem fail to hold, so that firms are not small relative to the total market, the competitive process of entry dynamics has to be modified. First of all, the entry blocking strategies or barriers to entry (BTE) have to be explicitly introduced. Defining net entry (NET) by the difference of entry over exit, it would now depend on BTE variables, time lags and new technology (τ):

$$\mathrm{d(NET)}/\mathrm{d}t = f(\mathrm{BTE}, t, \tau; \pi - \pi^0) \tag{4.5}$$

where π^0 is the average industry level profits greater than zero. Here the zero profit condition does not hold and it is replaced by the excess profit defined as $\pi - \pi^0$, which has to be positive (negative) to evoke potential entry (exit). The time lag in adjustment can be viewed in terms of a discrete time process

$$\mathrm{NET}_{t+1} - \mathrm{NET}_t = f(\mathrm{BTE}_t, \tau_t; \pi_t - \pi^0) \tag{4.6}$$

The composite BTE variable as a vector includes the entry blocking strategies of the Cournot–Nash players and it is assumed that the relative success in these strategies will increase the probability of retarding new entry. Thus increasing market power or dominance by BTE strategies is motivated by the expectation of higher future profits.

4.1.1 Genetic evolution and selection

In genetic models of species evolution entry is comparable to invading one species population by another. Game-theoretic models have been applied here to explain why some populations are successful in preventing invasion. The evolutionary strategies they follow help to maintain a stable population through gains in fitness. Fitness is comparable to economic efficiency in the economic theory of the firm. Thus fitness frontier is the economic efficiency frontier, which in dynamic context separates two groups of firms: one on the frontier and the other below. Models of two-person games are most appropriate here. It is useful to consider Cournot–Nash and also Pareto–Nash equilibrium solutions here. Such models generalize the Fisherian replicator dynamics by introducing the interactions in strategy by the two groups of players who compete in a fluctuating environment.

To be more specific we consider the limit pricing model of firm growth, where one firm is dominant in view of its position as a leader in leading-edge technology and the others are in the competitive fringe with less superior technology. This framework is typically one of Cournot–Nash equilibrium (CNE), where the dominant firm may use a predatory pricing strategy under some circumstances – so as to reduce or eliminate the market share of the rival firms. This type of behavior is comparable to the game-theory framework developed by Smith (1982) and others in their genetic models of evolution. In this game two species (hawk and dove) are contesting a resource of value v, which increases the Darwinian fitness of an individual. In the limit pricing game the dominant firm competes for the market share (or overall profit share) where the increased market share implies lower average cost through high minimum efficient scale (MES) levels of output. In the genetic fitness game, v is the gain in fitness to the winner, while c is the cost of an injury which reduces fitness. In the market game the cost c may represent the risk cost of fluctuations in profits or losses. Consider now the payoff matrix of this hawk (H) and dove (D) game in terms of changes in fitness as an illustration:

Strategy	H	D	$v = 2, c = 4$	
			H	D
H	$(v - c)/2$	v	-1	2
D	0	$v/2$	0	1

Three assumptions underlie the payoff matrix above of the symmetric game: (i) hawk versus hawk; each contestant has a 50% chance of injuring its opponent and obtaining the resource v and a 50% chance of being injured; (ii) hawk versus dove: hawk gets the resource and the dove retreats (exits) before being injured and (iii) dove versus dove: the resource is shared equally by the two contestants.

Now consider an infinite population of individuals, each adopting the strategy H or D at random (i.e., a mixed strategy). In this example due to Smith (1982) the behavior with higher fitness has an advantage in reproducing. Thus if the population consists only of doves, then this will continue in the next generation but if there is a mutation that introduces a hawk into the population, then over time the fraction of doves in the population will decrease and the fraction of hawks will increase. Similarly a mutant dove in a population of all hawks can successfully invade the population. The only *stable population* is thus evenly distributed between hawks and doves; here mutation would not disturb the population distribution. This defines the evolutionary stable strategy (ESS) strategy. As Mailath (1992) has shown, the ideas underlying the ESS is that a stable pattern of behavior in a population should be able to eliminate any invasion by a mutant. Thus if a population pattern of behavior is to eliminate invading mutations, it must have a higher fitness than the mutant. Suppose a population is playing a mixed strategy vector p and there is invasion by a group q with a mixed strategy vector q.

Let us perturb the strategy vector from p to $(1-\varepsilon)p+\varepsilon q$, where $0 \leq \varepsilon \leq 1$ specifies the influx. Then the strategy q will not succeed in invading p group, if p is fitter than q. This gives

$$E(p, (1-\varepsilon)p+\varepsilon q) > E(q, (1-\varepsilon)p+\varepsilon q) \tag{4.7}$$

where $E(\cdot)$ is the payoff to the mixed strategies of the two players. Now the condition (4.7) will hold for sufficiently smaller number ε if any only if

$$E(p,p) \geq E(q,p) \tag{4.8}$$

and if $E(p,p) = E(q,p)$, then $E(p,q) > E(q,q)$ the two parts of (4.8) have a simple interpretation as follows:

The first part says that if p is to resist invasion, it must do so at least as well as q against the p-group. The second part says that if p does exactly as well as q, then we have to look at the much less frequent encounters with the q-group. Thus the mixed strategy vector p is an ESS in Maynard Smith's definition if for all $q \neq p$ the two conditions of (4.8) hold. If we denote the matrix of fitness or payoff by $A = (a_{ij})$, then the fitness of the mixed strategy can be written in vector-matrix notation as $p'Aq = \sum_i \sum_j p_i a_{ij} q_j$, with prime denoting a row vector. Thus the conditions (4.8) can be written as

$$p'Ap \geq q'Ap$$
$$\text{and if} \quad p'Ap = q'Ap, \ \text{then } p'Aq > q'Aq,$$

defining an ESS strategy vector. ESS strategies can be viewed as a refinement of Nash equilibrium, since not all symmetric Nash equilibria are ESS.

One way to characterize the process of adjustment for reaching the ESS vector is the evolution dynamics as follows:

$$\mathrm{d}p_i/\mathrm{d}t = p_i(e_i'Ap - p'Ap) \tag{4.9}$$

where $e_i = 1$ for each $i = 1, 2, \ldots, n$. In this process the fraction playing strategy i is determined by how well i faces in terms of fitness relative to the population average fitness, measured by $(p'Ap)$. This replicator dynamics is useful in testing for stability of the evolution process. We note however that the notion of ESS is a rather narrow one for asymmetric contexts. Consider for example two populations in state $(p, q) \in S_n \times S_m$

where S_n and S_m are the two strategy sets. Clearly this state will not be stable in any evolutionary sense when there exists a state (x, y) near (p, q) such that both populations can increase their payoff (average fitness) by switching to it. More formally we may say that a pair (p, q) is a Nash–Pareto pair for an asymmetric game with payoff matrices A and B if the following two conditions hold:

(1) Equilibrium condition

$$p\acute{A}q \geq x'Aq \text{ and } q'Bp \geq y'Bp$$

for all $(x, y) \in S_n \times S_m$

(2) Stability condition

For all states $(x, y) \in S_n \times S_m$ for which the equality holds in (1) we have

if $x'Ay > p'Ay$, then $y'Bx < q'Bx$ and

if $y'Bx > q'Bx$, then $x'Ay < p'Ay$

The first condition says that (p, q) is a Nash equilibrium pair. The second condition says that it is impossible for both players to take simultaneous advantage from a deviation from the equilibrium pair (p, q), that is, at least one of them gets penalized. This corresponds to the concept of Pareto optimality. Thus we have an important result due to Hofbauer and Sigmund (1988):

Theorem 1. The pair (p, q) is a Nash–Pareto pair for the bimatrix game (A, B) if and only if there exists a positive constant c such that

$$(x - p)'Ay + c(y - q)'Bx \leq 0 \tag{4.10}$$

holds for all states (x, y) near (p, q). If (p, q) is a Nash–Pareto pair then the equality holds in (4.10) whenever the set sup(x) is contained in sup(p) and sup(y) in sup(q) respectively. Here sup$(\cdot)$ denotes the supremium. In particular if (p, q) is totally mixed, then

$$(x - p)'Ay + c(y - q)'Bx = 0$$

Two implications of this result are useful for the market games in economic theory. One is that if (p, q) is a Nash–Pareto pair, then it is stable for the standard replicator dynamics noted in (4.9). For example the dynamics is

$$\dot{x}_i = x_i(1 - x_i)(b - (a + b)y_j), \quad i \in S_m, \ j \in S_n$$
$$\dot{y}_j = y_j(1 - y_j)(d - (c + d)x_i), \quad i \in S_m, \ j \in S_n$$

defined on $[0, 1] \times [0, 1]$ with $a = (p - x)'Ay$, $b = (x - p)'Aq$, $c = (q - y)'Bx$ and $d = (y - q)'Bq$. The pair (p, q) is a Nash equilibrium pair if and only if $b \geq 0$ and $d \leq 0$ for all vectors x and y. It is stable against (x, y) in the sense that $p'Aq > x'Aq$ and $q'Bp > y'Bp$ for all $x \neq p$ and $y \neq q$. Thus we obtain the ESS definition for asymmetric games. It reduces to the notion of a strict Nash equilibrium.

For the introduction of any y-mutant satisfying $a > 0$ and $d < 0$ into the second population would drive back both populations to the pair (p, q). Thus the following concept of a *weak ESS* could still guarantee some evolutionary stability, that is, $b \leq 0$ with $a > 0$ if $b = 0$; and $d \leq 0$ with $c > 0$ if $d = 0$.

Due to the asymmetry of the game however there remains one possibility which has no analogue to the symmetric case. This is exactly the concept of Nash–Pareto pair: $b \leq 0$ and $d \leq 0$; if $b = 0 = d$, then $ac < 0$.

A second implication is that in symmetric games an evolutionary stable state is in general robust against small perturbations in the payoffs. In asymmetric conflicts the situation is different. Thus we have to define robustness slightly differently, for example, let (p, q) be a Nash–Pareto pair of a bimatrix game (A, B). It is said to be *robust*, if every game $(\overline{A}, \overline{B})$ in a suitable neighborhood of the given game (A, B) has a Nash–Pareto pair $(\bar{p}, \bar{q})$ near (p, q).

4.2 Dynamic interaction models

In this section we discuss replicator dynamics in terms of interactions of populations in fluctuating environments. This is comparable to an oligopolistic market structure, where at $t = 0$ $n \geq 2$ firms enter the industry. Each firm seeks to maximize the present discounted value of its profit stream over the horizon $[0, \infty]$ at a constant positive rate of interest r. We follow Stokey (1986) to characterize this noncooperative dynamic game in which the state variable is cumulative industry production denoted by x, the strategies are production decision rules and the payoffs are discounted profits. This model allows industry-wide learning by firms by assuming that unit costs $c_i(x)$ of each firm i declines when cumulative production x increases. Each firm i maximizes his payoff (profits)

$$\pi_i(g_1, g_2, \ldots, g_n; \hat{x}) = \int_0^{\infty} e^{-rt} g_i(x(t))[p\tilde{x}(t) - cx(t)]\,dt,$$

$$\text{where} \quad \tilde{x}(t) = \sum_{j=1}^{n} g_j(x(t)), \quad x(0) = \hat{x} \tag{4.11}$$

and $g_i(x)$ is the production rate of firm i when the cumulative industry output is $x(t)$. A sub game perfect Nash equilibrium is then defined by a vector of strategies $(g_1, \ldots, g_n)$ such that

$$\pi_i(g_1, \ldots, g_n; \hat{x}) \geq \pi_i(g_1, \ldots, g_{i-1}, g_i', g_{i+1}, \ldots, g_n; \hat{x}) \text{ for all } g_i'.$$

Stokey shows that there exists a unique symmetric Nash equilibrium in the space of production strategies for which the aggregate industry production is at least as great as that which a monopolist in a static environment with unit cost $c(x)$ would produce, but no more than the quantity at which price equals minimum unit cost $c(X_1)$, where X_1 is such that if $x \geq X_1$, then no further industry-wide learning occurs and from that point on the firms behave like ordinary Cournot competitors. Thus $g(x) = y_i(c(X_1)), x \geq X_1$ where $y_i(c)$ is defined by the reaction function (4.1) before.

Entry can be introduced in this framework in terms of a fixed cost of entry F. Let $v_i(x)$ be the present discounted value of the profit of each of n firms at the symmetric Cournot equilibrium. Clearly $v_i(x)$ is decreasing in n. Hence the number of firms in this industry will be given by N satisfying

$$v_N(X_1) \leq F < v_{N+1}(X_1) \tag{4.12}$$

where X_1 has been defined before in terms of the production rate $g(x) = y_i(c(X_1)), x \geq X_1$. This condition (4.12) says that the profits of the Nth firm would cover its cost of entry but those of the $(N+1)$th firm would not. Moreover since $v_N(x)$ is continuous and strictly increasing in x, the Nth firm enters when the cumulative industry experience is X_N satisfying $v_N(X_N) = F$.

Some simulation studies by Stokey show that changes in industry structure in this Cournot–Nash framework including changes in the market elasticity of demand have significant effects on the final equilibrium price and unit costs.

4.2.1 Departure from optimality: coexistence

Cournot–Nash type models are based on individual optimality of profits, when it is consistent with a reciprocal behavior by other players. This is typified by the solution of the simultaneous dynamic reaction curves one for each player. However in genetic models of evolution of species populations there occur many cases of departure from the mutual optimality consideration resulting in coexistence. It is useful to consider such cases, since they are applicable in many industrial and market situations where firms engage in networking and cooperating in joint research so as to exploit the economies due to large scale, indivisibility and industry-wide learning.

For growth of species in a stable environment the annual growth rate $\lambda = \lambda(x, \xi, N)$, that is, $\dot{x}/x = \lambda$ depends on the strategy chosen (ξ) in the actual environment and the actual density (N). Now the "best" strategy maximizing $\lambda(x, \xi, N)$ can be calculated for every N in a constant environment. This strategy has the following characteristics:

(a) it is an ESS, that is, its established population cannot be invaded by any other strategy;
(b) it can initially increase in the established population of any other strategy; and
(c) it increases not only initially but also it spreads until it becomes established excluding the former strategy.

But the situation changes when the environment (i.e., the overall economy) is not constant but fluctuating. In a population with fluctuating density the long-run growth of a strategy generally depends not only on a single but on several attributes of the density distribution, for example, mean, variance and higher moments.

We consider now two competing strategies with log growth rates

$$\ln \lambda_1(t) = -a(N(t) - K_1) - b(N(t) - K_1)^2 \tag{4.13}$$

$$\ln \lambda_2(t) = -c(N(t) - K_2)$$

Here $N(t)$ is total density, K_1, K_2 are the respective equilibrium densities of strategies 1 and 2 respectively in a stable environment. These strategies $\lambda_1(t)$, $\lambda_2(t)$ exploit a single limiting resource denoted by the density parameter N. Hence in a stable genetic environment they are not able to coexist: if $K_1 > K_2$ then strategy 1 excludes strategy 2, while $K_1 < K_2$ presents the reverse case. Now consider a random shock $\xi(t)$ acting on strategy 1 alone in (4.13):

$$\ln\lambda_1(t) = \xi(t) - a(N(t) - K_1) - b(N(t) - K_1)^2 \tag{4.14}$$
$$\ln\lambda_2(t) = -c(N(t) - K_2).$$

Here $\xi(t)$ is assumed to be a stationary zero mean process. The long-term behavior of the model (4.14) is then determined by the average growth rates $y_i = \overline{\ln\lambda_i(t)}$:

$$y_i = \overline{\xi}(t) - a(\overline{N} - K_1) - b(\overline{N} - K_1)^2 - bV(N) \tag{4.15}$$
$$y_2 = -c(\overline{N} - K_2)$$

where bar denotes average and $V(N)$ is variance of N. Note that strategy 2 is indirectly affected through density dependence.

Now coexistence is possible if both strategies have positive boundary growth rates (Δ_i) in the established population of the other strategy. Since the stationary solution of strategy 2 is $y_2 = \overline{\ln\lambda_2} = -c(\overline{N} - K_2) = 0$, that is, $\overline{N} = K_2$, the boundary growth rate of strategy 1, that is, Δ_1 is positive if

$$\Delta_1 = -a(K_2 - K_1) - b(K_2 - K_1)^2 > 0 \tag{4.16}$$

but $K_1 < (a/2b)$ since we require ln $\lambda_1(t)$ to be decreasing in $N(t) > 0$, hence we obtain

$$K_2 < K_1 \tag{4.17a}$$

The long-term stochastic equilibrium for strategy 1 is

$$y_1 = -a(\overline{N}_1 - K_1) - b(\overline{N}_1 - K_1)^2 - V(N) = 0.$$

In a stable environment $v(N) = 0$, hence $\overline{N} = K_1$. But in a fluctuating environment $V(N)$ is positive. This yields

$$\overline{N}_1 < K_1.$$

Now strategy 2 (species two) has a positive boundary growth rate if

$$\Delta_2 = -c(\overline{N}_1 - K_2) > 0 \tag{4.17b}$$

which holds if $\overline{N}_1 < K_2$

Thus we conclude from (4.17a) and (4.17b) that the two species (or strategies) are able to coexist if

$$\overline{N}_1 < K_2 < K_1$$

where $\overline{N}_1$ is the average density in the established population of species one.

This coexistence result depends on the quadratic form of $\ln \lambda_1(t)$ and a linear form of $\ln \lambda_2(t)$. An established population of strategy one determines $\overline{N} = K_2$ and $V(N) = 0$ and can be invaded by strategy 1 if $K_2 < K_1$ as in (4.17a) irrespective of the environmental fluctuation. The position of the population of strategy 1 however depends on the amount of environmental fluctuation: a weak fluctuation leads to a smaller $V(N)$. If fluctuation is strong enough then strategy 1 is located on the left side of the line $y_2 = 0$. In this case strategy 2 can invade the established population of strategy 1, while a population of strategy 2 remains invadable for strategy 1: the two strategies can coexist.

One implication of this result due to Levins (1979) is that if fluctuations are very weak, then the average density $\overline{N}_1$ of species one would be in the vicinity of the stable environmental density K_1. In this case there is only a very small interval $[\overline{N}_1, K_1]$ to choose K_2 so as to ensure coexistence. Thus the weaker the fluctuation (i.e., lower $V(N)$), the more restrictive are the requirements for coexistence.

Two economic implications of this genetic model of growth of two species or strategy-based population are of some importance. First, the limit pricing model of a dominant firm may be viewed as a coexistent market structure, where overall fluctuations in demand in the high-tech sector may be very strong as in computers and telecommunications. Second, product differentiations and intense product promotion strategies may affect the overall fluctuations of demand. The presence of indivisible fixed factors and the requirements of large scale may help to augment the fluctuations measured by the variance $V(N)$.

4.3 Models of differential games

The framework of differential games comes very naturally, when one attempts to model a market with $n \geq 2$ suppliers each with output, price or mixed strategies. The limit pricing model provides an important example, where a monopolist considers the possible threat of a new entry. Since the perception of potential entry is crucial to the entry dynamics one has to analyze the role of uncertainty and limited information available to the players to see how they affect the equilibrium solution of the differential game model. The conjectural variations and the perceived influence of each player on the market are the two most important aspects analyzed by Fershtman and Kamien (1985), who consider a dynamic market as

$$\dot{x} = f(x, u, t), \quad x(0) = x_0 \text{ given} \tag{4.18a}$$

with x and u as the state and control vectors and a dot denoting the time derivative. Each player chooses a component u_i of the control vector to maximize his payoff function appropriately discounted:

$$\max J_i = \int_0^\infty F_i(x, u, t)\,\mathrm{d}t, \quad i = 1, 2, \ldots, n \tag{4.18b}$$

Appropriate conditions of differentiability boundedness and concavity of the functions $F_i(\cdot)$ are assumed here to insure that the necessary conditions of optimality are also sufficient. Depending on the information structure available to each player, three types of strategy choices may be made:

$$\begin{aligned} u_i &= g_i(x_0, x, t): \text{ closed-loop no memory policy} \\ u_i &= g_i(x, t): \text{ feedback policy} \\ u_i &= g_i(x_0, t): \text{ open-loop policy} \end{aligned} \tag{4.18c}$$

Note that both closed-loop and feedback policies depend on the current state variable $x = x(t)$ and hence are easy to update with new information on $x(t)$. The open-loop policy however does not necessarily possess the feedback property along the optimal trajectory except when the control model is of the LQG (linear quadratic Gaussian) variety, that is, $F_i(\cdot)$ is quadratic, $f(\cdot)$ is linear and the additive error to $f(\cdot)$ is independently Gaussian or normal. We define a Nash equilibrium solution in a feedback form as an n-tuple decision rule $(g_1^*(x, t), \ldots, g_n^*(x, t))$ such that the following inequality holds

$$\begin{aligned} &J_i(g_1^*, \ldots, g_n^*) \geq J_i(g_1^*, \ldots, g_{i-1}^*, g_i, g_{i+1}^*, \ldots, g_n^*) \\ &\text{for all } i = 1, 2, \ldots, n \end{aligned} \tag{4.18d}$$

In other words g_i^* specifies the best response of player i to the strategies of the other players.

Now introduce the functions $\tilde{h}_i(h_1(x,t),\ldots,h_{i-1}(x,t),h_{i+1}(x,t),\ldots,h_n(x,t)$ as the conjectures of firm i about its rivals' behavior and assume that it satisfies some regularity conditions known as Lipschitz conditions, that is, continuity and boundedness. Then player i solves the optimal decision problem as

$$\begin{aligned} \max_{u_i} \quad & J_i = \int_0^\infty F_i(x,\tilde{h}(x,t),u_i)\,\mathrm{d}t \\ \text{s.t.} \quad & \dot{x}_i = f(x,\tilde{h}_i(x,t),u_i) \\ & x(0) = x_0 \text{ given} \end{aligned} \tag{4.19}$$

Let $\{\tilde{u}_i(t), 0 < t \leq \infty\}$ be the optimal solution trajectory, when it exists. Given the $(n-1)$ conjecture functions $\tilde{h}_i(x,t)$ and the optimal control path $\tilde{u}_i(t)$ one could compute the optimal path $\tilde{x}(t)$ of the state vector by using the equation of motion $\dot{x}_i = f(\cdot)$ in (4.19). Thus if the time path of the state variables $\tilde{x}(t)$ is available, then player i can easily construct the expected time path of control $(\tilde{u}_1(t),\tilde{u}_2(t),\ldots,\tilde{u}_n(t))$ which may be defined by

$$R_i(h) = R_i(h_1,h_2,\ldots,h_n) = R_i(\tilde{h}_i,\ldots,\tilde{h}_n)$$

In this framework Fershtman and Kamien define two types of equilibrium solutions as: a conjectural equilibrium (CE) and a perfect conjectural equilibrium (PCE). A CE is an n-type of conjectures $h^* = (h_1^*,\ldots,h_n^*)$ such that $R_i(h^*) = R_j(h^*)$ for every $i \neq j$. And a PCE is an equilibrium solution if it holds for all possible initial values (x_0, t_0). They then show that every conjectural equilibrium $(h_1^*(x,t),\ldots,h_n^*(x,t))$ constitutes a closed-loop no memory Nash equilibrium and vice versa.

As an example consider a two player game where market price p for a single good satisfies the equation of motion

$$\begin{aligned} &\dot{p} = s(a - b(u_1+u_2) - p^0) + \varepsilon \\ &p(0) = p_0 \text{ given}, \quad 0 < s < \infty; \ \varepsilon \sim \mathrm{NID}(0,1) \end{aligned}$$

where u_1, u_2 are the outputs. Assuming the cost functions to be quadratic $c(u_i) = cu_i + \frac{1}{2}u_i^2$, the payoff function is

$$\max_{u_i} J_i = E\int_0^\infty [\mathrm{e}^{-rt}(pu_i - cu_i - \tfrac{1}{2}u_i^2)]\,\mathrm{d}t,$$

with E as expectation and ε is independent Gaussian. Since this is in the form of an LQG model, the optimal feedback strategies $u_i^*(t)$ exist for each player as:

$$u_i^*(t) = [1 - bsk(t)]p(t) + bsm(t) - c$$

$$\text{where} \quad k(t) = (6s^2b^2)^{-1}\left[(r + 4bs + 2s) - \left\{(r + 4bs + 25)^2 - 12s^2b^2\right\}^{1/2}\right]$$

$$m(t) = (r - 3b^2s^2k(t) + s + 2bs)^{-1}[c - ask - 2bsck(t)].$$

It is clear that the PCE price path exists by LQG theorem and converges to a unique steady state price p^* where

$$p^* = [2b(1 - bsk(t) + 1)^{-1}\{a + 2b(c - bsm(t))\}].$$

Several comments may be added here. First of all, one has to note that even when CE conjectures $h^* = (h_1^*, \ldots, h_n^*)$ fail to exist, there may exist mixed strategy solutions which may be introduced by the tracing approach of Harsanyi (1975). Secondly, the Cournot–Nash solution of a noncooperative game can be improved in many cases by recognizing that the two players are interdependent. The reaction theory of reaching a Cournot–Nash equilibrium solution in a two-person game explicitly allows an adaptive trial and error process before reaching equilibrium, where one player takes the strategy of the other player as a parameter. However this parameter has to be estimated each time the player makes his move. Such estimates can in may cases help them realize that they can improve the noncooperative solution and these may be formed through "public" information available to each player or through "private" information subjectively held by each player (e.g., Aumann's (1974) subjective random devices). Correlation of strategies here introduced through such information channels may improve profits of both firms, that is, Pareto improvement. In many modern high-tech industries like computers, telecommunications, such limited forms of cooperation and correlation of R&D strategies have been increasingly adopted through networking and related devices.

4.3.1 Learning-by-doing and scale economies

From an economic viewpoint an important barrier to entry comes from the dynamic scale economies through investment in new technology by the incumbent and established firms. We have already discussed the industry-wide learning case in the dynamic model due to Stokey (1986), where each firm's average cost declines due to this learning effect. We now consider leaning-by-doing through cumulative experience affecting the rate of production. In this setup it is important to distinguish between the current output $\dot{y}(t) = dy/dt$ and cumulative output $y(t)$ in the production function. Assume now Cournot-type duopolistic markets with each producer producing differentiated outputs $\dot{y}_i(t) = dy_i/dt$ and selling at prices $p_1(t)$ and $p_2(t)$. The first one is an established firm maximizing its discounted profit function

$$\begin{aligned} \max \quad & J_1 = \int_0^\infty \exp(-rt)[p_1(t) - c_1(t)\dot{y}_1]\,dt \\ \text{s.t.} \quad & \dot{y}_i = f_i(y_1, y_2; p_1, p_2); \quad i = 1, 2 \end{aligned} \tag{4.20}$$

On using the current value Hamiltonian $H = \exp(-rt)[(p_1 - c_1 + \lambda_1)f_1 + (p_2 - c_2 + \lambda_2)f_2]$ and assuming the standard regularity conditions for the existence of an optimal trajectory, Pontryagin's maximum principle specifies the following necessary conditions for optimality for $i = 1, 2$:

$$\begin{aligned} &\dot{\lambda}_i = r\lambda_i - \partial H/\partial y_i \\ &\dot{y}_i = f_i(y_1, y_2, p_1, p_2); \quad \partial H/\partial p_i = 0, \quad \text{all } t \\ &\text{and} \quad \lim_{t\to\infty} \exp(-rt)\lambda_i(t) = 0. \end{aligned}$$

On using $\mu_{ii} = (\partial f_i/\partial p_i)(p_i/f_i)$ and $\mu_{ji} = (\partial f_j/\partial p_i)(p_i/f_j), j \neq i$ as own price elasticity and cross elasticity of demand the optimal price rule for the established firm can be written as

$$p_1 = (1 + \mu_{11})^{-1}\,[\mu_{11}(c_1 - \lambda_1) - \lambda_2\mu_{21}],$$

with the optimal trajectory for $\lambda_1(t)$ as

$$\dot{\lambda}_1(t) = r\lambda_1 + (c_1 - \lambda_1 - p_1)(\partial f_1/\partial y_1) + f_1(\partial c_1/\partial y_1) \tag{4.21}$$

where $\mu_{11} < 0$ and $\mu_{21} > 0$ for rival products and $\partial c_1/\partial y_1 < 0$, i.e., unit cost declines due to learning in the form of cumulative output.

The optimal pricing rule (4.21) shows the price to be much lower than the monopoly price $(1 + \mu_{11})^{-1}(\mu_{11}(c_1 - \lambda_1))$ for every positive value of $\lambda_2(t)$. Second, future cost declines, that is, $\partial c_1/\partial y_1 < 0$ tends to reduce the dynamic shadow price $\lambda_1(t)$. Moreover, if the demand function depends only on the prices $\dot{y}_i = f(p_1, p_2)$ alone, then the sign of $\dot{p}_1$ may be shown as

$$\text{sign}(\dot{p}_1) = \text{sign}(-r\lambda_1 - (\partial c_1/\partial t) - \lambda_2\mu_{21}).$$

This shows a strong pressure for price declines over time. Intensive competition has forced this type of pressure in the semiconductor and R&D-intensive industries in the Silicon Valley today.

The pressure on price declines arises basically through the decline in unit costs. For an industry comprising n firms a specific firm i grows in market share s_i if

$$\dot{s}_i/s_i = \lambda(\bar{c} - c_i), \quad i = 1, 2, \ldots, n \tag{4.22a}$$

it is characterized by below the industry-average unit cost ($\bar{c}$) where $\bar{c} = \sum_{i=1}^{n} c_i s_i$. If s_i becomes very low then it is assumed to exit, that is, no new firms enter. The parameter λ is the speed of selection, that is, the speed at which firm market shares react to differences between firm efficiency characteristics. The equation (4.22a) yields

$$\dot{s}_1 = \lambda s_1(1 - s_1)(c_2 - c_1)$$

where the cases $c_2 - c_1 = g(s_2 - s_1)$, $\partial g/\partial x > 0$, $x = s_2 - s_1$ and $c_2 - c_2 = f(s_2 - s_2)$, $\partial f/\partial x < 0$, $x = s_2 - s_1$ comprise decreasing and increasing returns to scale respectively. Mazzucato (2000) has incorporated both increasing (IRS) and decreasing returns (DRS) to scale dynamics in a single polynomial cost equation

$$c_2 - c_1 = (1 - 2s_1)(s_1 - a)(s_1 - b) \tag{4.22b}$$

which embodies IRS between market share 0 and $1/n$ and between $1/n$ and 1 and DRS within b and a. This equation (4.22b) yields

$$\dot{s}_1 = \lambda s_1(1 - s_1)(1 - 2s_1)(s_1 - a)(s_1 - b) \quad \text{with } 0.5 < a < 1 \text{ and } 0 < b < 0.5.$$

The points b and a are unstable equilibrium points here, since very small disturbances will cause divergence away from these points whereas the point $1/n$ is a stable equilibrium for small disturbances. In the limit the firm with IRS tends to dominate the market.

Thus by using this entry dynamics model (4.22a) the interactions between the established firm and the new entrants can be directly analyzed. The exit behavior may also be modeled in terms of total outputs y_1 and y_2. The growth paths of the two sectors may be represented as

$$d \ln y_1/dt = a_1(d_1 - (y_1 + y_2))$$

$$d \ln y_2/dt = a_2(d_2 - (y_1 + y_2))$$

Here d_1, d_2 are the maximum demand attainable by the two firms through their advertising strategies when $D = d_1 + d_2$ in total demand. Let $\alpha = a_1/a_2$. Then

$$\dot{y}_1/y_1 - \alpha\dot{y}_2/y_2 = d\ln(y_1/y_2^{\alpha})/\mathrm{d}t = a_1(d_1 - d_2).$$

Clearly if $d_1 > d_2$, the ratio of y_1 to y_2^{α} and thus the ratio of y_1 to y_2 will increase without limit and hence the entrant group will go extinct.

Thus we conclude that Cournot–Nash selection processes are essentially based on several entry-preventing strategies such as increasing returns to scale, learning and intensified rate of advertising. However the usual Cournot–Nash reaction function approach does not include strategies by which both players could gain, for example, Pareto optimal strategies. We may consider such a case in a Cournot–Nash market model, where by increasing the volume of total market demand through simultaneous advertising expenditure each player may attain Pareto improvement.

To be specific we assume two payoffs of two players as total cost functions J_1, J_2 each of which is convex in both strategies u_1, u_2. Since by lowering costs the firm can increase its market share, we choose cost as payoff function. Pareto frontier's are obtained by minimizing with respect to u_1, u_2 the convex function $(1 - w)J_1 + wJ_2$ for $0 < w < 1$, which is a weighted combination. Necessary and sufficient conditions for Pareto optimality and Nash equilibrium are given below:

$$\text{Pareto:} \quad (1 - w)(\partial J_1/\partial u_1) + w(\partial J_2/\partial u_1) = 0$$

$$(1 - w)(\partial J_1/\partial u_2) + w(\partial J_2/\partial u_2) = 0$$

$$\text{Nash:} \quad \partial J_1/\partial u_1 = 0, \partial J_2/\partial u_2 = 0.$$

It follows that a Nash equilibrium belongs to the Pareto efficiency frontier if and only if

$$\partial J_1/\partial u_1 = \partial J_1/\partial u_2 = \partial J_2/\partial u_1 = \partial J_2/\partial u_2 = 0.$$

As this condition is independent of w, if a Pareto efficient point is an equilibrium point, then all efficient points are Nash equilibrium points. Conversely if a Pareto efficient point dominates a Nash equilibrium point, not all points of the Pareto frontier dominate this point. In other words the values of w corresponding to points on the Pareto frontier that dominate the Nash equilibrium point belong to some closed interval [0,1].

An example may clarify this point. Assume that payoff functions in terms of quadratic costs

$$J_1(u) = (1/2)u'Au + u'a + c_1$$
$$J_2(u) = (1/2)u'Bu + u'b + c_2$$

where $u = (u_1, u_2)$ is the strategy vector, prime denotes a row vector and A, B are symmetric positive definite matrices. Necessary conditions for Pareto optimality and Cournot–Nash equilibrium then become

$$\text{Pareto}: \quad [(1-w)A + wB]u^P = -(1-w)a - wb;$$
$$\text{Nash}: \quad \begin{bmatrix} A_{11} & A_{12} \\ B_{21} & B_{22} \end{bmatrix} u^N = -\begin{pmatrix} a_1 \\ b_2 \end{pmatrix}$$

where u^P and u^N are Pareto and Nash equilibrium vectors. These necessary conditions are sufficient since each J_1, J_2 are convex. We also require the matrix associated with the column vector u^N to be nonsingular for the uniqueness of Nash equilibrium. Under uniqueness the Nash equilibrium holds if and only if $Au^N = -a$ and $Bu^N = -b$. Clearly this occurs when $J_2 = rJ_1 + s$ and in particular for $r = -1$ and $s = 0$ (zero-sum case).

4.4 Concluding remarks

Game-theoretic selection processes provide a rich variety of entry and exit dynamics in market competition. This is so because firms can apply a variety of entry preventing barriers followed by diverse payoff functions in noncooperative game-theory framework. Sometimes the incentives to cooperate may yield Pareto optimal solutions through an increase in total market demand or total industry-wide R&D investments.

There exist several differences of the Cournot–Nash adjustment process from the Walrasian competitive adjustment and the stability–instability characteristics of the time-path so adjustments are sometimes different. Differences in scale economies affecting unit costs, stochastic shocks affecting the cost efficiency differences between the incumbent and established firms and the weaker speed of selection in the Cournot–Nash process of adjustment are some of the major sources of difference between the Cournot–Nash and Walrasian adjustment.

References

Aumann, R.J. (1974): Subjectivity and correlation in randomized strategies. *Journal of Mathematical Economics* 1, 67–96.

Dasgupta, P. and Stiglitz, J. (1980): Industrial structure and the nature of innovative activity. *Economic Journal* 90, 249–65.

Harsanyi, J.C. (1975): Solution concepts in game theory. International journal of game theory 5, 39–54.

Hofbauer, J. and Sigmund, K. (1988): *The Theory of Evolution and Dynamical Systems: Mathematical Aspects of Selection*. Cambridge University Press, New York.

Levins, R. (1979): Coexistence in a Variable Environment. *Nature* 114, 765–83.

Mailath. G.J. (1992): Introduction to the symposium on evolutionary game theory. *Journal of Economic Theory* 57, 259–77.

Mazzucato, M. (2000): *Firm Size, Innovation and Market Structure*. Edward Elgar, Cheltenham.

Novshek, W. (1980): Cournot Equilibrium with free entry. *Review of Economic Studies* 47, 473–86.

Smith, J.M. (1982): *Evolution and the Theory of Games*. Cambridge University Press, Cambridge.

Stokey, N.L. (1986): The dynamics of industry-wide learning. In: Hell, W., Starr, R. and Starrett, D. (eds), *Equilibrium Analysis*. Cambridge University Press, New York.

5
Innovations and Growth

In the modern technology-intensive industries of today investments in R&D and knowledge capital have played a crucial role as engines of growth. The evolution of industry has been profoundly affected by innovations in product design, process of production and software development, which have helped some firms to enhance their core competence and managerial efficiency. This has also intensified the pressure of competition, which has been called a state of *hypercompetition* by the experts in industrial organization theory and management science.

R&D investments have several dynamic features of the innovations process, which affect the industry evolution process. These features have significant economic implications for the selection and adjustment processes underlying the competitive and oligopolistic market structures. Our objective here is twofold: to discuss the different implications of R&D investment as it affects the competitive process of entry and exit in the evolution of industries and to relate innovations to the competitive edge of core competence by the leading firms. We discuss some theoretical models which incorporate spillovers of cooperative and non-cooperative R&D expenditures and their impact on employment growth and industry evolution. Since R&D expenditure usually involves increasing returns to scale (IRS), it directly affects the costs of entry and exit. We consider here a dynamic model of entry, where the evolution of the number of firms in terms of entry and exit are endogenously determined, that is, the change in the number of firms is assumed to adjust up to the point where the costs of entry (of new firms) or exit (of existing firms) equal the net present value of entry. This type of model is closely related to the dynamic limit pricing model, where new firms are assumed to enter the industry as soon as the market price tends to exceed the limit price. R&D expenditures by some firms may reduce the limit price through reduction of unit costs through internal or external economies of scale. We consider some theoretical extensions of these dynamic models of R&D investment as it affects the equilibrium dynamics involving different number of firms and different R&D investment levels, which may imply multiple equilibria due to the nonconvexity of the underlying cost functions.

5.1 R&D investment and innovation efficiency

Innovations take many forms but in a broad sense they involve developing new processes, new products and new organizational improvements. R&D investment plays an active role in innovations in new processes and in new products and services. Several dynamic features of R&D investment by firms are important for selection and industry evolution. First of all, R&D expenditure not only generates new knowledge and information about new technical processes and products, but also enhances the firm's ability to assimilate, exploit and improve existing information and hence existing "knowledge capital". Enhancing this ability to assimilate and improve exiting information affects the learning process within an industry that has cumulative impact on the industry evolution. For example, Cohen and Levinthal (1989) have argued that one of the main reasons firms invested in R&D in semiconductor industry is because it provides an in-house technical capability that could keep these firms on the leading edge of the latest technology and thereby facilitate the assimilation of new technology developed elsewhere.

A second aspect of R&D investment within a firm is its spillover effect within an industry. R&D yields externalities in the sense that knowledge acquired in one firm spills over to other firms and very often knowledge spread in this way finds new applications both locally and globally and thereby stimulates further innovative activity in other firms.

Finally, the possibility of implicit or explicit collaboration in R&D networking or joint ventures increases the incentive of firms to invest. This may encourage more industry R&D investment in equilibrium. In the absence of collaboration the competing firms may not invest enough, since innovations cannot be appropriated by the inventor, for example, his competitors will copy the invention and thus "free ride" without paying for it. Thus the basic reason for the success of joint R&D ventures is that externalities or spillovers are internalized thus eliminating free rides.

We consider first the empirical basis of R&D innovations in modern industries and then its implications for selection and industry evolution. Cohen and Levinthal (1989) have made an important contribution in this area by analyzing the two faces of R&D investment in terms of spillover and externality. One impact of R&D spillovers emphasized by Nelson, Arrow and others is that they diminish firm's incentive to invest in R&D and the related production. The other impact discussed by Cohen and Levinthal emphasizes the point that spillovers may encourage equilibrium industry R&D investment, since the firm's R&D investment develops its ability to exploit knowledge from the environment, that is, develops its "absorptive" capacity or learning by which a firm can acquire outside knowledge. Thus a significant benefit of a firm's R&D investment is its contribution to the intra-industry knowledge base and learning, by which externality and spillovers may help firms develop new products and/or new processes.

The model developed by Cohen and Levinthal (C&L) starts with the firm's stock of knowledge and denotes the addition to firm's stock of technological and scientific knowledge by z_i and assumes that z_i increases the firm's gross earnings π^i but at a diminishing rate. The relationship determining z_i is assumed to be of the form

$$z_i = M_i + \gamma_i \left(\theta \sum_{j \neq i} M_j + T \right), \quad 0 \leq \gamma_i \leq 1 \tag{5.1}$$

where M_i is the firm's R&D investment, γ_i is the fraction of intra-industry knowledge that the firm is able to exploit, θ is the degree of intra-industry spillover of research knowledge. M_j represents other firms' ($j \neq i$) R&D investments which contribute to z_i and θ denotes the degree to which the research effort of one firm may spill over to a pool of knowledge potentially available to all other firms, for example, $\theta = 1$ implies that all the benefits of one firm's research accrue to the research pool potentially available to all other firms, whereas $\theta = 0$ implies that the research benefits are exclusively appropriated by the firm conducting the research.

It is assumed that $\gamma_i = \gamma_i(M_i, \beta)$ depends on both M_i (the firm's R&D) and β, where β is a composite variable reflecting the characteristics of outside knowledge, that is, its complexity, ease of transferability and its link with the existing industry-specific knowledge. Clearly the composite variable β will differ from one industry to another, for example, in pharmaceutical industry it may involve lot of experimentation, long gestation periods and the complexity of the marketing process for new drugs, whereas for the computer industry it may involve software experimentation and the ease of application in multiple situations. It is assumed that the composite variable β denoting "ease of learning" is such that a higher level indicates that the firm's ability to assimilate outside knowledge is more dependent on its own R&D expenditure. Thus it is assumed that increasing β increases the marginal effect of R&D on the firm's absorptive capacity but diminishes the level of absorptive capacity.

C&L evaluate the effects of increasing the explanatory variables such as β, θ and T on the equilibrium value of firm's R&D investment denoted by M^*, where it is derived from maximizing π^i with respect to M_i as

$$R = \mathrm{MC} = 1 \tag{5.2}$$

where MC is the marginal cost of R&D expenditure equal to one and R is marginal return given by

$$R = \pi^i_{z_i}\left[1 + \gamma_{M_i}\left(\theta\sum_{j\neq i} M_j + T\right)\right] + \theta\sum_{j\neq i}\gamma_j\pi^i_{z_j} \tag{5.3}$$

where the subscripts denote partial derivatives. On solving the equations (5.2) and (5.3) simultaneously one obtains the equilibrium value of each firm's R&D denoted by M^*.

The impact on M^* of the explanatory variables β, θ and T are derived as:

$$\mathrm{sign}(\partial M^*/\partial\beta) = \mathrm{sign}\left[\pi^i_{z_i}\{\gamma_{M\beta}(\theta(n-1)M + T) + \theta(n-1)\frac{\partial\gamma}{\partial\beta}\pi^i_{z_i}\right] \tag{5.4}$$

$$\mathrm{sign}(\partial M^*/\partial\beta) = \mathrm{sign}\left[\pi^i_{z_i}\gamma_M(n-1)M + (n-1)\gamma\pi^i_{z_j}\right] \tag{5.5}$$

and

$$\mathrm{sign}(\partial M^*/\partial T) = \mathrm{sign}\left[\gamma_M\pi^i_{z_i} + \left(\pi^i_{z_iz_j} + (n-1)\pi^i_{z_iz_j}\gamma(1+\gamma_M T)\right]\right. \tag{5.6}$$

The first term on the right hand side of (5.4) shows that a higher β induces the firm with more incentives to conduct R&D, because its own R&D has become more critical to assimilating its rivals' spillovers $\theta(n-1)M$ and the extra-industry knowledge T. The second term shows a decline in rivals' absorptive capacity $(n-1)\gamma$ as β increases. As a result the rival competitors are less able to exploit the firm's spillover. Due to both these effects the payoffs to the firm's R&D increases and *ceteris paribus* more R&D investment is induced.

The effect of θ on M^* is ambiguous, due to two offsetting effects: the benefit to the firm of increasing its absorptive capacity denoted by the first term and the loss associated with the diminished appropriability of rents denoted by the second term on the right-hand side of (5.5). Note however that the desire to assimilate knowledge generated by other firms provides a positive incentive to invest in R&D as θ increases.

The relation (5.5) shows that with an endogenous absorptive capacity, the firm has a positive incentive to invest in R&D to exploit the pool of external knowledge. With $\gamma_M = 0$, that is, zero endogenous absorptive capacity the sign $(\partial M^*/\partial T)$ is negative, since a higher T merely substitutes for the firm's own R&D, that is, $\pi^i_{z_i z_j} < 0$.

C&L estimate by regression (OLS, GLS and Tobit) models the effects of the knowledge inputs and other industry characteristics on unit R&D expenditure (intensity) of business units. The sample data included R&D performing business units consisting of 1 302 units representing 297 firms in 151 lines of business in the US manufacturing sector over the period 1975–77. The empirical data were obtained from FTC's (Federal Trade Commission) Line of Business Program and the survey data collected by Levin *et al.* (1987). A set of estimates of selected regression coefficients is reproduced in Table 5.1. Appropriability here is defined as follows: the respondents in Levin *et al.* (1987) survey were asked to rate on a 7-point scale the effectiveness of different methods used by firms to protect the competitive advantages of new products and new processes. For a line of business appropriability is then defined as the maximum score. Thus if appropriability increases the spillover level declines and hence R&D intensity increases. The new plant variable is used to reflect the relative maturity of an industry's technology, that is, it measures the percentage of an industry's plant and equipment installed during the five years preceding 1977 as reported in FTC's dataset. Industry demand conditions are represented by the price and income elasticity measures.

The explanatory variables T and β are measured indirectly for the survey data. The level of extra-industry knowledge T is measured by five sources of which three are reported in Table 5.1, for example, downstream users of industry's products (usertech), government agencies and research laboratories (govtech) and university research (univtech). The proxy variables used for β in Table 5.1 represent cumulativeness and the targeted quality of knowledge which are all field-specific, hence research in basic and applied sciences is reported here, for example, the characteristic that distinguishes the basic from the applied sciences is the degree to which research results are targeted to the specific needs of firms, where basic science is less targeted than the applied. Hence the β value

Table 5.1 Effects of knowledge and other explanatory variables on R&D intensity

	OLS	GLS	Tobit
1. *Technological opportunity*			
(a) appropriability $(1-\theta)$	0.396* (0.156)	0.360** (0.104)	0.260 (0.161)
(b) usertech	0.387** (0.99)	0.409** (0.070)	0.510** (0.166)
(c) univtech	0.346** (0.128)	0.245** (0.089)	0.321* (0.147)
(d) govtech	0.252* (0.100)	0.170* (0.076)	0.200* (0.100)
2. *Basic science research*			
(a) biology	0.176 (0.096)	0.042 (0.057)	0.159 (0.116)
(b) chemistry	0.195** (0.071)	0.095 (0.050)	0.149 (0.078)
(c) physics	0.189 (0.109)	0.037 (0.082)	0.156 (0.109)
3. *Applied science research*			
(a) computer science	0.336** (0.123)	0.157 (0.093)	0.446** (0.121)
(b) agricultural science	−0.373** (0.084)	−0.253** (0.055)	−0.259* (0.101)
(c) materials science	−0.005 (0.121)	−0.028 (0.089)	0.231* (0.116)
4. *New plant*	0.055** (0.008)	0.041** (0.006)	0.042** (0.007)
5. *Elasticity of*			
(a) price	−0.180** (0.061)	−0.071 (0.044)	−0.147* (0.060)
(b) income	1.062** (0.170)	0.638** (0.136)	1.145** (0.180)
R^2	0.278	—	—

Notes: Only a selected set of regression coefficient estimates are given here with standard errors in parentheses. One and two asterisks denote significant values of *t* tests at 5% and 1% respectively.

associated with basic science research is higher than that associated with applied science. As a result the coefficient values of the technological opportunity variables associated with the basic sciences should exceed those of the applied sciences. The estimates in Table 5.1 show that except computer science the coefficients are uniformly greater for the basic sciences. The exception of computer science may also be due to the rapid advance in software and process development, where the basic and applied knowledge are intermingled.

Some broad conclusions emerge from the estimates reported in Table 5.1. First of all, the results reject across all three estimation methods the hypothesis that the effects on R&D spending of the basic and applied science are equal. This means that the role of learning differs significantly across fields in terms of cumulativeness, targetedness and the pace of advance which affect the influence of technological opportunity on R&D spending. Second, increasing technological opportunity through the less targeted basic sciences evokes more R&D spending than does increasing the technological opportunity through applied sciences. Finally, the OLS and GLS estimates of the coefficient of appropriability is positive and significant, implying that spillovers have a net negative effect on R&D. However if we consider the coefficient estimates of the interaction between appropriability and price elasticity of demand the results are as follows:

	OLS	**GLS**	**Tobit**
Interaction variable	−0.192 (0.106)	−0.200* (0.091)	−0.176 (0.103)

This suggests the existence of positive absorption incentives associated with spillovers.

Next we consider an application in the computer industry, which has witnessed rapid technological changes in recent years in both hardware and software R&D. Recent empirical studies have found cost-reducing effects of rapid technological progress to be substantial in most technology-intensive industries of today such as microelectronics, telecommunication and computers. Two types of productivity growth are associated with such technological progress: the scale economies effect and the shift of the production and cost frontiers. Also there exist substantial improvements in the quality of inputs and outputs. The contribution of R&D expenditure has played a significant role here. This role involves learning in different forms that help improve productive efficiency of firms. One may classify learning into two broad types: one associated with technological and the other with human capital. Three types of measures of learning are generally used in the literature. One is the cumulative experience embodied in cumulative output. The second measure is cumulative experience embodied in strategic inputs such as R&D investments in Arrow's learning by doing models. Finally, the experience in "knowledge capital" available to a firm due to spillover from other firms may be embodied in the cost function through the research inputs.

Unlike the regression approach of Cohen and Levinthal we now develop and apply a nonparametric and semiparametric model of production and cost efficiency involving R&D expenditure and its learning effects. These nonparametric models do not use any specific form of the cost or production function; they are based on the observed levels of inputs, outputs and their growth over time. Technological progress (regress) is measured in this framework by the proportional rate of growth (decline) of total factor productivity (TFP), where TFP is defined as the ratio of aggregate (weighted) output to aggregate (weighted) inputs. The nonparametric efficiency model is specified here in terms of a series of linear programming (LP) models, the unifying theme of which is a convex hull method of characterizing the production frontier without using any market prices (i.e., technical efficiency) and the cost frontier (i.e., allocative efficiency which utilizes the input prices).

Consider now a standard input oriented nonparametric model, also known as a DEA (data envelopment analysis) model for testing the relative efficiency of a reference firm or decision making unit $h(\text{DMU}_h)$ in a cluster of N units, where each DMU_j produces s outputs (y_{rj}) with two types of inputs: m physical inputs (x_{ij}) and n R&D inputs as knowledge capital (z_{wj}):

$$\text{Min} \quad \theta + \phi, \tag{5.7}$$

$$\text{s.t.} \quad \sum_{j=1}^{N} X_j\lambda_j \leq \theta X_h; \quad \sum_{j=1}^{N} Z_j\lambda_j \leq \phi Z_h$$

$$\sum_{j=1}^{N} Y_j\lambda_j \geq Y_h; \quad \sum_{j} \lambda_j \geq 0; \quad j = 1, 2, \ldots, N$$

Here X_j, Z_j and Y_j are the observed input and output vectors for each DMU_j, where $j = 1, 2, \ldots, N$. Let $\lambda^* = \left(\lambda_j^*\right), \theta^*, \phi^*$ be the optimal solutions of model (5.7) with all slacks zero. Then the reference unit or firm h is said to be *technically efficient* if $\theta^* = 1.0 = \phi^*$. If however θ^* and ϕ^* are positive but less than unity, then it is not technically efficient at the 100% level, since it uses excess inputs measured by $(1 - \theta^*)x_{ih}$ and $(1 - \phi^*)z_{wh}$. Overall efficiency (OE_j) of a unit j however combines with technical (TE_j) or production efficiency and the allocative (AE_j) or price efficiency as follows: $\text{OE}_j = \text{TE}_j \times \text{AE}_j$. To measure overall efficiency of a DMU_h one solves the cost minimizing model:

$$\text{Min} \quad C = c'x + q'z \tag{5.8}$$

$$\text{s.t.} \quad X\lambda \leq x; \quad Z\lambda \leq z; \quad Y\lambda \geq Y_h; \quad \lambda'e = 1; \quad \lambda \geq 0$$

where e is a column vector with N elements each of which is unity, prime denotes transpose, c and q are unit cost vectors of the two types of inputs x and z which are now the decision variables and $X = (X_j), Z = (Z_j)$ and $Y = (Y_j)$ are appropriate matrices of observed inputs and outputs. Denoting optimal values by asterisks, technical efficiency is

$\text{TE}_h = \theta^* + \phi^*$ as before, overall efficiency (OE_h) is C_h^*/C_h computed from model (5.8) and hence the allocative efficiency is $\text{AE}_h = C_h^*/(\theta^* + \phi^*)C_h$, where C_h and C_h^* are the observed and optimal costs for unit h.

Now consider the special characteristics of the research inputs z. Since these inputs lower the initial unit production costs c_i and also affect the cost function nonlinearly we can rewrite the objective function of (5.2) as

$$\text{Min TC} = \sum_i \left[\left(c_i - f_i \left(\sum_w q_w z_w \right) \right) x_i + \frac{1}{2} d_i x_i^2 \right] + \frac{1}{2} \sum_{w=1}^{n} g_w z_w^2 \tag{5.9}$$

subject to the constraints of model (5.2). Here f_i is the unit cost reduction with $f_i < c_i$ and the component cost functions are assumed to be strictly convex implying diminishing return to the underlying R&D production function. The optimal solutions z_w, x_i and λ_i now must satisfy the Kuhn–Tucker conditions as follows:

$$\begin{aligned} &f_i q_w x_i + \gamma_w \le g_w z_w; \quad z_w \ge 0 \\ &f_i \left(\sum q_w z_w \right) + \beta_i \le c_i + d_i x_i; \quad x_i \ge 0 \end{aligned} \tag{5.10}$$

If the unit (DMU_h) is efficient with positive input levels and zero slacks, then we must have equality $\partial L/\partial z_w = 0 = \partial L/\partial x_i$ where L is the Lagrangean function. Hence we can write the optimal values (z_w^*, x_i^*) as:

$$\begin{aligned} z_w^* &= \left(f_i q_w x_i^* + \gamma_w^* \right) / g_w; \quad w = 1, 2, \ldots, n \\ x_i^* &= \left(g_w z_w^* - \gamma_w^* \right) / (f_i q_w); \quad i = 1, 2, \ldots, m \end{aligned} \tag{5.11}$$

By duality the production frontier for unit $j = 1, 2, \ldots, N$ satisfies

$$\alpha^{*\prime} Y_j \le \alpha_0^* + \beta^{*\prime} X_j + \gamma^{*\prime} Z_j; \quad (\alpha^*, \beta^*, \gamma^*) \ge 0$$

where the equality holds if unit j is efficient and there is no degeneracy due to congestion costs. Clearly a negative (positive or zero) value of α_0^* implies increasing (diminishing constant) returns to scale.

Note that this generalized quadratic programming model (5.9) has many flexible features. First of all, if the research inputs are viewed as cumulative stream of past investment as in Arrow model of learning-by-doing, then the cost function TC in (5.9) may be viewed as a long-run cost function. Given the capital input z^* the reference firm solves for the optimal current inputs x_i^* through minimizing the short-run cost function $\mathrm{TC}(x|z^*)$. Second, the learning effect parameter $f_i > 0$ shows that the efficiency estimates through DEA model (5.8) would be biased if it ignores the learning parameters. Third, the complementarity (i.e., interdependence) of the two types of inputs is clearly brought out in the linear relation between x_i^* and z_w in (5.11). For example, it shows that

$$\partial x_i^*/\partial z_w^* > 0, \qquad \partial x_i^*/\partial f_i > 0, \qquad \partial x_i^*/\partial \beta_i^* > 0$$

and

$$\partial z_w^*/\partial x_i^* > 0, \qquad \partial z_w^*/\partial f_i > 0, \qquad \partial z_w^*/\partial \gamma_w^* > 0$$

Finally, compared to a linear program this quadratic programming model (5.9) permits more substitution among the inputs, thus making it possible for more units to be efficient.

One limitation of the long-run cost (5.3) minimization model above is that it ignores the time profile of output generated by cumulative investment experience. Let $z(t) = (z_w(t))$ be the vector of gross investments and $k(t) = \int_0^t z(s)\,ds$ be the cumulative value where

$$\dot{k}_w(t) = z_w(t) - \delta_w k_w(t) \tag{5.12}$$

δ_w : fixed rate of depreciation

In this case the transformed DEA model becomes dynamic as follows:

$$\text{Min} \int_0^\infty e^{-\rho t}[c'(t)x(t) + C(z(t))]\,dt$$

subject to (5.12) and the constraints of model (5.8)

Here $C(z(t))$ is a scalar adjustment cost, which is generally assumed nonlinear in the theory of investment. This type of formulation has been recently analyzed by Sengupta (2003), which shows the stability and adaptivity aspects of convergence to the optimal path.

Another type of characterization of the research inputs and their productivity is in the current literature of new growth theory. Thus Lucas (1993) considered a growth process where each firm has a production function, where its output depends on its own labor and physical capital inputs as well as the total knowledge capital of the whole industry. The availability of industry's knowledge capital occurs through the spillover mechanism or diffusion of the underlying information process. The utilization of the industry's knowledge capital by each firm has been called by Jovanovic (1997) as the learning effect which is very significant in the modern software based industries. To characterize this learning effect we introduce a composite input vector X_j^C for DMU_j as the share of each DMU_j out of the industry total supply of each input, for example, $\sum_{j=1}^{N} X_{ij}^C = X_i^T$, where X_i^T is the total industry supply of input i. We can then formalize the input-oriented DEA model in two forms as before:

$$\begin{aligned}
\text{Min} \quad & \theta + \phi \\
\text{s.t.} \quad & \sum_{j=1}^{N} X_j \lambda_j \leq \theta X_h; \quad \sum_j X_j^C \lambda_j \leq \phi X_h^C \\
& \sum_j Y_j \lambda_j \geq Y_h; \quad \lambda' e = 1, \quad \lambda \geq 0
\end{aligned}$$

and

$$\begin{aligned}
\text{Min} \quad & C = c'x + q'x^C \\
\text{s.t.} \quad & X\lambda \leq x; \quad X^C \lambda \leq x^C; \quad Y\lambda \geq Y_h; \quad \lambda' e = 1; \quad \lambda \geq 0
\end{aligned} \tag{5.13}$$

On using the Lagrange multipliers α, β, γ and α_0, the production frontier for DMU_h may be easily derived from the dual problem as

$$\alpha' Y_h = \beta' X_h + \gamma' X_h^C + \alpha_0; \quad \alpha, \beta, \gamma \geq 0$$

Clearly the interdependence of the two inputs x and x^C can be easily introduced in this framework through nonlinear interaction terms in the objective function of (5.13) or, through the method used in (5.9) before.

Thus the generalized DEA models incorporate three additional sources of relative efficiency not found in the conventional DEA models: (1) unit cost reduction due to the complementarity effect of R&D inputs, (2) the increasing returns to scale due to learning by doing and finally (3) the spillover effect of knowledge capital in the industry as a whole.

Consider now a dynamic cost frontier with knowledge capital. For this purpose it is useful to aggregate all inputs into a single cost and relate this cost to a single output in order to derive a cost frontier. Like a dynamic production frontier, a dynamic cost frontier helps to characterize the technological progress associated with the input and output growth.

Consider the production function of a firm with a single output (y) and m inputs (x_i) at time t

$$y = f(x_1, x_2, \ldots, x_m, t) \tag{5.14}$$

Let $\dot{A}/A = (\partial f/\partial t)(1/f)$ denote the proportional shift in the production function with time and it is called technological progress (regress) which we wish to relate to the index of productivity. On taking the time derivative of equation (5.14) and dividing by output one obtains

$$\dot{y}/y = \sum_{i=1}^{m} \frac{\partial f}{\partial x_i} \frac{x_i}{y} \frac{\dot{x}_i}{x_i} + \dot{A}/A$$

where dot denotes the time derivative. Assume that the firm minimizes the total cost (c) for producing y. Then the first order conditions for cost minimization imply $\partial f/\partial x_i = w_i/(\partial c/\partial y)$ where $\partial c/\partial y$ is marginal cost and w_i is the input price of x_i. On using this result in (5.2) one obtains

$$\dot{y}/y = \sum_{i=1}^{m} \varepsilon^{-1} \frac{w_i x_i}{c} \frac{\dot{x}_i}{x_i} + \dot{A}/A \tag{5.15}$$

where $\varepsilon = (\partial c/\partial y)(y/c)$ is the elasticity of cost with respect to output and $c = \sum w_i x_i$ is total cost of inputs. To aggregate inputs into a single composite input (F) we follow the Divisia index method of aggregation due to Diewert (1976), Denny *et al.* (1981) and others where

$$\dot{F}/F = \sum_{i=1}^{m} \frac{w_i x_i}{c} \frac{\dot{x}_i}{x_i} \tag{5.16}$$

On combining (5.15) and (5.16) one obtains

$$\dot{A}/A = \dot{y}/y - \varepsilon^{-1}(\dot{F}/F).$$

Since $\text{TFP} = y/F$, the proportional rate of growth of TFP is

$$\frac{\dot{\text{TFP}}}{\text{TFP}} = \frac{\dot{y}}{y} - \frac{\dot{F}}{F}.$$

Also it follows

$$\dot{A}/A = \frac{\dot{\text{TFP}}}{\text{TFP}} + (1 - \varepsilon^{-1})\frac{\dot{F}}{F}$$

or

$$\frac{\dot{\text{TFP}}}{\text{TFP}} = \frac{\dot{A}}{A} + (\varepsilon^{-1} - 1)\frac{\dot{F}}{F}.$$

Clearly if there is constant returns to scale then $\varepsilon = 1$ and hence

$$\dot{A}/A = \dot{\text{TFP}}/\text{TFP}$$

Otherwise the TFP growth rate exceeds (falls short of) the growth rate of A when there exists increasing (diminishing) returns to scale.

By duality the shift in the production function can be easily related to the shift in the total cost function

$$c = g(w_1, w_2, \ldots, w_m, t)$$

Let $\dot{B}/B = (1/c)(\partial g/\partial t)$ be the proportionate shift in the cost function (5.14). Then one can show that

$$(-\dot{B}/B) = \frac{\dot{\text{TFP}}}{\text{TFP}} + (\varepsilon - 1)\frac{\dot{y}}{y} \tag{5.17}$$

or

$$\frac{\dot{\text{TFP}}}{\text{TFP}} = -\frac{\dot{B}}{B} + (1 - \varepsilon)\frac{\dot{y}}{y} \tag{5.18}$$

Denny *et al.* (1981) have applied this model to estimate the growth of TFP in Canadian telecommunications industry by assuming a translog cost function in a regression framework. It is clear from (5.18) that in case of constant returns to scale ($\varepsilon = 1$) we have

$$\dot{\text{TFP}}/\text{TFP} = -\dot{B}/B = \dot{A}/A,$$

that is, technological progress ($\dot{A}/A$) equals the downward shift ($-\dot{B}/B$) of the cost function. Moreover if the production function (5.9) is log-linear, then it can be specified as

$$\ln c = \ln B + \sum_{i=1}^{m} \tilde{\beta}_i \ln w_i + \left(1/\sum_{i=1}^{m} \beta_i\right) \ln y$$

where $\tilde{\beta}_i = \beta_i / \sum_{i=1}^{m} \beta_i$ and β_i is the elasticity of output with respect to input x_i in the log-linear production function. Clearly the cost elasticity of output is given by $\varepsilon = \left(\sum \beta_i\right)^{-1}$, where the case of IRS yields scale economies, that is, $\varepsilon < 1.0$.

When input price data are available, these can be utilized in defining an aggregate input F_j for firm j by following the Divisia index formulation (5.13) before. Let $\hat{F}_j = \Delta F_j / F_j$ and $\hat{y}_j = \Delta y_j / y_j$ denote the proportional growth rates of input and output for firm j. Then one can test the relative efficiency of firm h by setting up the LP model

$$\text{Min} \quad \phi \tag{5.19}$$

$$\text{s.t.} \quad \sum_{j=1}^{N} \hat{F}_j \mu_j \leq \phi \hat{F}_h; \quad \sum_j \hat{y}_j \mu_j \geq \hat{y}_h$$

$$\sum \mu_j = 1, \mu_j \geq 0; \quad j = 1, 2, \ldots, N$$

If firm h is efficient (i.e., $\phi^* = 1.0$ with all slacks zero), then this yields the simple production frontier

$$\Delta y_h / y_h = \frac{b^*}{a^*} \frac{\Delta F_h}{F_h} + \frac{b_0^*}{a^*}$$

where the Lagrangean function is

$$L = -\phi + b\left(\phi \hat{F}_h - \sum \hat{F}_j \mu_j\right) + a\left(\sum \hat{y}_j \mu_j - \hat{y}_h\right) + b_0\left(1 - \sum \mu_j\right)$$

with ϕ, b, a nonnegative and b_0 unrestricted in sign. On comparing (5.13) and (5.22) it is easy to deduce that the cost elasticity of output for firm h is $\varepsilon_h^* = a^*/b^*$ and $\Delta A_h / A_h = b_0^* / a^*$, where asterisk denotes optimal values. Hence the downward shift of the cost frontier for the efficient firm h can be specified as

$$-\Delta B_h / B_h = \frac{b_0^*}{a^*} + \left(\frac{b^*}{a^*} - 1\right) \frac{\Delta F_h}{F_h}$$

Clearly if $b^* = a^*$ then $\Delta A_h / A_h = -\Delta B_h / B_h$, that is, the upward shift of the production frontier equals the downward shift of the cost frontier for the efficient firm h.

Note that by varying h within the set $I_N = \{1, 2, \ldots, N\}$ the industry can be decomposed into two groups of firms: one containing the relatively efficient firms and the other comprising nonefficient ones. Two implications of this cost-oriented formulation are important for analyzing the dynamics of a technology-intensive industry such as computers or semiconductors. One is the exit behavior of firms which are not growth efficient, that is, high cost firms fail to survive the competitive pressure. On replacing F_j by its associated cost C_j, the cost frontier for the efficient firm h can also be written as

$$\Delta C_h/C_h = (a^*/b^*)(\Delta y_h/y_h) - b_0^*/b^*$$

When firms are not on this dynamic cost frontier above, they get crushed by competitive pressure in the long run, that is, they are more vulnerable to being squeezed out of the market.

A second aspect of this dynamic model is that some types of nonlinearity can be easily built into it. This nonlinearity may reflect second order effects of output growth on the pattern of input and cost growth. Thus consider the LP model (5.21) with $\hat{F}_j = \hat{C}_j = \Delta C_j/C_j$ and adjoin the nonlinear constraint

$$\sum \hat{y}_j^2 \mu_j = \hat{y}_h^2$$

as an equality so that its Lagrange multiplier α is unrestricted in sign. The cost frontier for the efficient firm h then gets transformed as follows

$$\Delta C_h/C_h = (a^*/b^*)(\Delta y_h/y_h) + (\alpha^*/b^*)(\Delta y_h/y_h)^2 - (b_0^*/b^*).$$

Clearly if α^* is negative, then as output growth rises, the marginal rate of cost falls, that is,

$$\partial \hat{C}_h/\partial \hat{y}_h = (a^*/b^*) + 2\alpha^* \hat{y}_h/b^* < 0$$

$$\text{if } \hat{y}_h = \Delta y_h/y_h > \frac{-a^*}{2\alpha}$$

In this case the efficient firms would tend to reap the benefits of declining costs at an increasing rate, thus forcing the inefficient firms to exit in the long run.

Next we consider the impact of the input called knowledge capital, which may exhibit the learning phenomena. Let k_j denote the knowledge capital for firm j and c_j be the average cost per unit of output. The efficient firm h seeks to minimize the discounted stream of average production costs in order to determine the optimal levels of costs (inputs) and knowledge capital, when investments (z) lead to the growth of knowledge capital. The decision model then takes the form

$$\begin{aligned}
\text{Min} \quad & J = \int_0^{\infty} e^{\rho t}(c + i(z))\,dt \\
\text{s.t.} \quad & \dot{k} = z(t) - \delta k \\
& \sum_{j=1}^{N} c_j\lambda_j \le c; \quad \sum_j \lambda_j y_j \ge y_h \\
& \sum_j \lambda_j y_j^2 = y_h^2, \quad \sum k_j\lambda_j \le k \\
& \sum \lambda_j = 1, \quad \lambda_j \ge 0; \quad j = 1, 2, \ldots, N
\end{aligned}$$

Here dot over a variable denotes the time derivative, $i(z)$ is the cost of investment and δ is the fixed rate of depreciation. Here the nonlinear output constraint $\sum \lambda_j y_j^2 = y_h^2$ is specified in equality form, so that its Lagrange multiplier (a) may be free in sign. Here firm h is tested for overall Pareto efficiency relative to the cluster of N firms in the industry. This type of dynamic model can be easily solved by Pontryagin's maximum principle, where one introduces the Hamiltonian function H as

$$H = e^{-\rho t}[c + i(z) + s(z - \delta k(t))]$$

where $s = s(t)$ is the adjoint function. If the optimal path of knowledge capital $k(t)$ exists, then by Pontryagin's maximum principle there must exist a continuous function $s(t)$ satisfying

$$\dot{s}(t) = (\rho + \delta)s - b.$$

Assuming interior optimal solutions the efficient firm h must satisfy the following necessary conditions which are also sufficient here:

$$\begin{aligned}
& \beta - 1 = 0, \text{that is, } \beta = 1 \\
& c_h = \beta_0 + a y_h^2 - b k_h; \quad \beta_0 \text{ and a free in sign.}
\end{aligned}$$

This dynamic cost frontier has several economic implications. First of all, the cost frontier in (5.25) for the efficient firm shows a decline in average production cost when the level of knowledge capital increases. Furthermore, if the coefficient a is negative, then for higher levels of output $y_h > \alpha/(2|a|)$ the marginal cost may also decline. Secondly, the steady state solution $(\bar{k}, \bar{s})$ on the optimal trajectory would be stable if the following two conditions hold

$$\bar{s}/u \lesseqgtr \delta\bar{k} \text{ according as } k \gtreqless \bar{k}$$

and

$$(\rho+\delta)\bar{s} \lesseqgtr \bar{b} \text{ according as } s \gtreqless \bar{s}$$

since the two differential equations above are linear. Also one could combine the two differential equations, that is, adjoint equations into a single second order equation

$$u\ddot{k}(t) - u\rho\dot{k}(t) - u\delta(\rho+\delta)k(t) + b = 0.$$

Its characteristic equation is

$$\mu^2 - \rho\mu - \delta(\rho+\delta) = 0$$

which shows the two roots to be real and opposite in sign, that is, $\mu_1 > 0$, $\mu_2 < 0$. Thus the steady state pair $(\bar{k}, \bar{s})$ has the saddle point property. Finally, if the observed path of accumulation of knowledge capital equals the optimal path over time, then the firm would exhibit dynamic efficiency; otherwise the conditional cost function would exhibit myopic inefficiency.

Now we consider an empirical application characterizing the role of R&D expenditure. An empirical application to a set of 12 firms (companies) in the US computer industry over a 12-year period (1987–98) is used here to illustrate the concept of dynamic efficiency. The selection of 12 companies is made from a larger set of 40 companies over a 16-year period. R&D input is used here as a proxy for knowledge capital. We have used Standard and Poor's Compustat Database (SIC codes 3570 and 3571) for the input and output data for the 12-year period 1987–98. This selection is based on two considerations: (1) survival of firms through the whole period and (2) promising current profit records. The companies are: Compaq, Datapoint, Dell, Sequent, Data General, Hewlett-Packard, Hitachi, Toshiba, Apple, Maxwell, Silicon Graphics and Sun Microsystems.

The single output variable (y) is net sales in dollars per year and the nine inputs in dollars per year are combined into three composite inputs: x_1 for R&D expenditure, x_2 for net plant and equipment and x_3 for total manufacturing and marketing costs.

We consider input-based level and growth efficiency in terms of non-radial measures of efficiency, that is, efficiency specific to each inputs. The level efficiency model is

$$\begin{aligned} \text{Min} \quad & \sum_{i=1}^{3} \theta_i \\ \text{s.t.} \quad & \sum_{j=1}^{12} y_j \lambda_j \geq y_k; \quad \sum_{j=1}^{12} x_{ij} \lambda_j \leq \theta_i x_{ik}, \qquad i = 1, 2, 3 \\ & \sum \lambda_j = 1, \quad \lambda_j \geq 0; \quad 0 \leq \theta_i \leq 1 \end{aligned} \tag{5.20}$$

and it is applied for three years 1987, 1991 and 1998. The growth efficiency model is used in the form

$$\begin{aligned} \text{Min} \quad & \sum_{i=1}^{3} \phi_i \\ \text{s.t.} \quad & \sum_{j=1}^{12} (\Delta y_j / y_j) \mu_j \geq \Delta y_k / y_k \\ & \sum_{j} (\Delta x_{ij} / x_{ij}) \mu_j \leq \phi_i (\Delta x_{ik} / x_{ik}), \quad i = 1, 2, 3 \\ & \sum \mu_j = 1, \quad \mu_j \geq 0; \quad 0 \leq \phi_i \leq 1 \end{aligned} \tag{5.21}$$

over 3-year periods 1987–90, 1991–94 and 1995–98. Tables 5.2 and 5.3 present the empirical results. It is interesting to observe that a firm which is sequentially level efficient over the three years 1987, 1991 and 1998 is not necessarily growth efficient. Such an example is Hitachi, whose net sales did not grow as fast as some of its competitors.

Measuring technological progress by the shadow price of the constraint $\sum \mu_j = 1$ the ranking of the 12 firms appears as follows:

Rank	Firms		
	1987–90	1991–94	1995–98
1.	Datapoint	Dell	Datapoint
2.	Data General	Silicon	Maxwell
3.	Sequent		Compaq
4.	HP		Dell
5.	Toshiba		Data General
6.	Sun		Sequent
7.	Hitachi		Silicon
8.	Apple		
9.	Compaq		
10.	Silicon		

Note that during 1991–94 only Dell and Silicon Graphics exhibited positive technological progress, others with zero or negative technological progress. Datapoint regained its first rank during the recent period 1995–98. Table 5.2 also shows Datapoint to be growth-efficient in terms of all the three inputs over all the periods. On writing the dual formulation of the LP model (5.20) the dynamic production frontier for the

Table 5.2 Non-radial measures (θ_i^*) of level efficiency

	1987			1991			1998		
	x_1	x_2	x_3	x_1	x_2	x_3	x_1	x_2	x_3
Data General	0.55	0.15	1.0	1.0	1.0	1.0	0.59	0.58	1.0
HP	1.0	1.0	1.0	1.0	1.0	1.0	1.0	1.0	1.0
Hitachi	1.0	1.0	1.0	1.0	1.0	1.0	1.0	1.0	1.0
Toshiba	1.0	1.0	1.0	1.0	0.96	0.97	1.0	1.0	1.0
Apple	1.0	1.0	1.0	1.0	1.0	1.0	0.40	0.69	1.0
Compaq	1.0	1.0	1.0	1.0	1.0	1.0	1.0	1.0	1.0
Datapoint	0.62	0.20	0.97	1.0	1.0	1.0	1.0	1.0	1.0
Dell	1.0	1.0	1.0	1.0	1.0	0.94	1.0	1.0	1.0
Maxwell	1.0	1.0	1.0	1.0	1.0	1.0	1.0	1.0	1.0
Sequent	1.0	1.0	1.0	1.0	1.0	1.0	0.64	0.53	1.0
Silicon Graphics	0.38	0.52	0.70	1.0	1.0	1.0	0.51	0.71	1.0
Sun Microsystems	0.57	0.29	1.0	1.0	1.0	1.0	1.0	1.0	1.0

Notes: x_1 represents R&D cost, x_2 represents Net plant and equipment cost and x_3 represents Total production cost.

Table 5.3 Non-radial measures (θ_i^*) of growth efficiency

	1987–90			1991–94			1995–98		
	x_1	x_2	x_3	x_1	x_2	x_3	x_1	x_2	x_3
Data General	1.0	1.0	1.0	1.0	1.0	1.0	0.48	0.54	0.77
HP	0.49	1.0	0.47	0.55	0.80	1.0	1.0	1.0	1.0
Hitachi	0.40	0.68	0.65	1.0	1.0	1.0	0.94	0.84	1.0
Toshiba	0.29	0.62	0.72	1.0	1.0	1.0	1.0	1.0	1.0
Apple	0.52	0.69	0.64	0.51	0.44	0.76	1.0	1.0	1.0
Compaq	0.40	0.54	0.50	1.0	1.0	1.0	0.33	0.60	0.75
Data Point	1.0	1.0	1.0	1.0	1.0	1.0	1.0	1.0	1.0
Dell	0.61	0.44	0.47	1.0	1.0	1.0	1.0	1.0	1.0
Maxwell	1.0	1.0	1.0	1.0	1.0	1.0	1.0	1.0	1.0
Sequent	1.0	1.0	1.0	1.0	1.0	1.0	0.50	0.54	0.48
Silicon Graphics	1.0	1.0	1.0	1.0	1.0	1.0	0.25	0.25	0.38
Sun Microsystems	1.0	1.0	1.0	1.0	1.0	1.0	0.42	0.24	0.67

efficient firm k may be written as

$$\alpha^*(\Delta y_k/y_k) = \beta_0^* + \sum_{i=1}^{3} \beta_i^*(\Delta x_{ik}/x_{ik}) \tag{5.22}$$

The values of the technological progress $(\beta_0^* > 0)$ or regress $(\beta_0^* < 0)$ appear as follows:

Firms	**Values**
1987–90	
Datapoint	10.74
Data General	3.32
Sequent	3.24
Toshiba	0.261
Sun	0.260
Hitachi	0.25
Apple	0.08
Compaq	0.040
Silicon	0.041
1995–98	
Datapoint	19.35
Maxwell	7.36
Compaq	1.86
Sequent	0.38
Silicon	0.10

Note that when one divides both sides of (5.31) by the optimal values of $\alpha*$, all the coefficients of input growth become less than one. The empirical details of this analysis have been analyzed by Sengupta (2004) elsewhere.

Measuring Solow-type technological progress by the ratio β_0^*/α^* we obtain the following results:

(i) Technological progress is above 4% per year for 58% of firms in 1987–90, 83% in 1991–94 and 75% in 1995–98.
(ii) Some typical examples are (in %):
  1987–90 HPC (6.4), DGC (4.9), DEL (3.8) and SUN (4.9)
  1991–94 DGC (5.1), HPC (9.0), APL (9.1) and SIL (6.4)
  1995–98 DGC (6.1), HIT (9.0), COM (16.4) and MAX (7.7)

The sum of the input coefficients $\sum_{i=1}^{3} \beta_i$ measuring returns to scale is equal to or exceeds unity in all cases of the sample of efficient firms.

A second way of analyzing the empirical results is to run a regression of the dependent variable log output $= \hat{y}$ on the three independent variables: log R&D ($\hat{x}_1$), log plant and equipment ($\hat{x}_2$) and log cost of goods sold ($\hat{x}_3$) with a dummy variable D for each coefficient, where $D = 1.0$ for the efficient firms and zero otherwise. The results are as follows:

$$\begin{aligned}
\text{1987–98} \quad \hat{y} &= 1.199^{**} + 0.162^{**}\hat{x}_1 + 0.065^{*}\mathrm{D}\hat{x}_1 \\
&\quad + 0.009\hat{x}_2 - 0.034\mathrm{D}\hat{x}_2 \\
&\quad + 0.743^{**}\hat{x}_3 + 0.034^{*}\mathrm{D}\hat{x}_3 \quad (R^2 = 0.996) \\
\text{1991} \quad \hat{y} &= 1.214^{**} + 0.262^{**}\hat{x}_1 - 0.075\hat{x}_2 \\
&\quad + 0.791^{**}\hat{x}_3 \quad (R^2 = 0.998)
\end{aligned}$$

D significant for $\hat{x}_1$ and $\hat{x}_3$ only

$$\begin{aligned}
\text{1998} \quad \hat{y} &= 0.925^{**} + 0.140^{*}\hat{x}_1 + 0.015\hat{x}_2 \\
&\quad + 0.0842^{**}\hat{x}_3 \quad (R^2 = 0.998)
\end{aligned}$$

D significant for $\hat{x}_1$ and $\hat{x}_3$ only

Clearly R&D expenditures have played a most dynamic role in the productivity growth of the efficient firms in the computer industry and this trend is likely to continue in the future.

To summarize we note that the role of knowledge capital and its cost reducing effects are incorporated here in order to capture learning-by-doing. The recent upsurge in the personal computer industry is illustrated here in terms of the two types of efficiency models: one deals with the level efficiency and the other the growth efficiency. Solow-type technological progress is estimated by the growth efficiency model applied to the computer industry over the period 1987–98. Two important findings are that all the growth-efficient firms exhibited technological progress on or above 4% per year on the average and the R&D input is significantly more important for the efficient firms than the nonefficient ones.

Now we consider the impact of own R&D expenditure and compare it with the spillover effect for the companies in the computer industry. Denoting proportional changes by a hat let $\hat{c}_j$, $\hat{y}_j$, $\hat{x}_{1j}$ and $\hat{x}_{2j}$ indicate changes in average costs, output, own R&D expenditure and the aggregate industry R&D spending available for firm j, then the dynamic efficiency model is used in the form

$$\begin{aligned} &\text{Min} \quad \theta \\ &\text{s.t.} \quad \sum_{j=1}^{N} \hat{c}_j\lambda_j \le \theta\hat{c}_h; \quad \sum_{j=1}^{N} \hat{y}_j\lambda_j \ge \hat{y}_h \\ &\qquad \sum_{j=1}^{N} \hat{x}_{1j}\lambda_j \le \hat{x}_{ih}; \quad \sum_{j=1}^{N} \hat{x}_{2j}\lambda_j \le \hat{x}_{2h} \\ &\qquad \sum_{j=1}^{N} \lambda_j = 1, \quad \lambda_j \ge 0; \quad j = 1, 2, \ldots, N \end{aligned} \tag{5.23}$$

Here $x_{2j} = \sum_{\substack{k\neq j \\ k=1}}^{N} x_{ik}$ is used as a proxy for industry-wide R&D knowledge capital. This is analogous to the measure M_j used by Cohen and Levinthal indicating other firms' R&D expenditure.

This model (5.23) is empirically applied to the computer industry data from Compustat over the period 1985–2000. The initial dataset for 40 companies was reduced to 12 due to heterogeneity and continuous availability consideration. Thus the analyzed data set consist mostly of hardware based companies, since the constraint of homogeneity is considered important for picking companies in the portfolio. Perfect homogeneity would exist, if each company and its products could be perfect substitutes for another company. However our selected data set only represents broad homogeneity and not perfect homogeneity.

The detailed characteristics of the empirical dataset have been analyzed by Sengupta (2003, 2004). Using net sales as output average cost excluding R&D expenditure is computed from total cost per unit of output for each company. Standard and Poor's Compustat data are used for calculating total cost as the sum of the following major cost components: R&D expenditure, net plant and equipment and total manufacturing and marketing costs as in (5.20) before.

The 16-year period 1985–2000 is divided into three subperiods 1985–90, 1990–95, 1995–2000 and 1997–2000, in order to convey the trend in terms of moving averages. It is clear from the LP model (5.22) that for a dynamically efficient firm j we would have θ close to unity and the cost frontier would appear as

$$\Delta c_j/c_j = \beta_0 - \beta_1(\Delta y_j/y_j) - \beta_2(\Delta x_{1j}/x_{1j}) - \beta_3(\Delta x_{2j}/x_{2j}), \qquad (5.24)$$
$$\beta_1, \beta_2, \beta_2 \geq 0, \beta_0 \text{ free in sign}$$

where β_i's are appropriate optimal values of Lagrange multipliers. Here β_2 and β_3 measure the impact of own and outside R&D spending respectively.

The average estimated values of β_2 and β_3 for selected efficient companies are as follows:

	1985–90		1990–95		1995–2000	
	β_2	β_3	β_2	β_3	β_2	β_3
Apple	1.23	0.35	1.26	0.31	0.94	0.02
Dell	2.49	0.47	0.74	0.05	1.10	0.31
HP	1.80	0.38	1.03	0.65	0.94	0.22
Silicon Graphics	0.91	0.19	0.95	0.21	1.42	0.75
Sun	1.05	0.24	0.85	0.25	1.01	0.23

If we run (5.24) as a regression model for 12 firms over the periods 1985–92 and 1992–2000 the results appear as follows:

	β_0	β_1	β_2	β_3	R^2
1985–92	−0.01*	0.004	1.25*	0.25	0.96
1992–2000	−0.02**	0.01**	1.41*	0.31*	0.92

Here one and two asterisks denote significance of t-values at 5% and 1% respectively. Two points emerge very clearly from these growth efficiency results. One is that the impact of R&D growth is highly significant;

this is particularly so for the leading firms. Since data are not separately available for hardware and software specific R&D expenditure, it is not possible to estimate their separate impact. But since average computer prices have followed a continuous downtrend, it may be reasonable to assume that this downtrend has been largely cost-driven. Second, the spillover effect has consistently followed the cost-reducing tendency, although at a lesser rate than the overall R&D spending. Recent outsourcing of R&D spending abroad may also intensify this tendency of spillover effect to be more and more important.

5.2 Core competence and industry evolution

What makes a firm grow? What causes an industry to evolve and progress? From a broad standpoint two types of answers have been offered. One is managerial, the other economic. The managerial perspective is based on organization theory, which focuses on cost competence as the primary source of growth. The economic perspective emphasizes productivity and efficiency as the basic source of growth. Economic efficiency of both physical and human capital including innovations through R&D have been stressed by the modern theory of endogenous growth.

Core competence rather than market power has been identified by Prahalad and Hamel (1994) as the basic cornerstone of success in the modern hypercompetitive world. Core competence has been defined as the collective learning of the organization especially learning how to coordinate diverse production skills and integrate multiple streams of technologies. Four basic elements of core competence are: learn from own and outside research, coordinate, integrate so as to reduce unit costs and innovate so as to gain market share through price and cost reductions.

A company's own R&D expenditures help reduce its long-run unit costs, also yield externalities in the sense of spillover. These spillovers may yield aggregate increasing returns to scale to R&D as the empirical models of Cohen and Levinthal (1989) and Sengupta (2004) discussed earlier in the section showed. We now consider a dynamic model due to Folster and Trofimov (1997), which relates this scale effect due to R&D investment and externalities to the evolution of firms through entry and exit dynamics. We generalize this model in terms of core competence and innovation efficiency in a Schumpeterian framework. The latter framework emphasizes a process of creative destruction as a basic of competitive and comparative advantage for the leading firm.

Consider an industry comprising n homogenous firms, where a representative incumbent has an expected net profit as $\pi = \pi(n, u)$, where n is the total number of firms and u is an R&D effort variable at time t allocated to product innovation. Let $u(n)$ denote the interior maximum of π with respect to u. Each incumbent firm chooses the time-path $u = u(t)$ of R&D that maximizes the present value $v_0 = \int_0^\infty \exp(-rt)\pi(n, u)dt$ of future profits at the real discount rate r. The current value of profits at time t is $v(t) = \int_t^\infty \exp(-r(\tau - t))\pi(n, u)d\tau$. Let z be the opportunity cost of entry into the industry and $v = v(t)$ the market value of the incumbent firm already defined. Then it is assumed that firms enter the industry when $v > z$, that is, when the net present value of entry $(v - z)$ is positive and exit from the industry when $v < z$. It is assumed that the cost of entry and exit is a linear function of the absolute change of the number of firms $\alpha^{-1}(n)$ where α is a positive parameter. The entry–exit dynamic is then of the form

$$\dot{n} = \alpha(v - z) \tag{5.25a}$$

and

$$\dot{v} = rv - \pi(n, u) \tag{5.25b}$$

where dot denotes time derivative and the equation (5.25b) is derived from the differentiation of $v(t)$ at time t.

The dynamic optimization decision for the firm is how to choose the time path of R&D effort u so that it maximizes the initial value $v_0 = v(0)$ subject to the dynamic equations (5.25a) and (5.25b). On using the Hamiltonian H defined as

$$H = \pi(n, u) + \theta_1(v - z) + \theta_2(rv - \pi(n, u))$$

where θ_1, θ_2 are costate variables and applying Pontryagin's maximum principle we obtain the following necessary conditions for the optimal path of $u = u(t)$:

$$\begin{aligned} \dot{\theta}_1 &= r\theta_1 - (1 - \theta_2)\pi'(n) \\ \dot{\theta}_2 &= -\alpha\theta_1 \end{aligned} \tag{5.26}$$

where $\pi(n) = \pi(n, u(n))$ and $\pi'(n)$ is $\partial\pi(n)/\partial n$. Also for each t the H function is maximized for the optimal control u. If $\theta_2 < 1$ then the optimal control $u(n)$ is identical with that which maximizes the static profit $\pi(n,u)$ with respect to u. If $\theta_1 > 0$ for all t, then (5.26) yields $\dot{\theta}_2 < 0$ and this implies that $\theta_2 < 1$. However the adjoint variable θ_1 is the shadow price of entry by new firm, which maybe usually positive if its initial value is chosen appropriately.

The major implications of the dynamic model (5.26) above are as follows. First of all, when $v = z$ in the steady state, then there is no entry or exit ($\dot{n} = 0$) and the steady state number of firms can be computed from the equation $\pi(n) = rz$. Second, one may linearize the system (5.26) and consider the reduced homogenous system as

$$\dot{n} - \alpha v = 0, \quad \dot{v} + \pi'(n) - rv = 0$$

The characteristic equation of this system is

$$\lambda^2 - r\lambda + \alpha\pi'(n) = 0$$

with two eigenvalues

$$\lambda_1, \lambda_2 = [(r \pm (r^2 - 4\alpha\pi'(n))^{1/2}]/2.$$

Two cases are important. When $\pi'(n^*) < 0$ then the eigenvalues are real and opposite in sign, thus implying that the stationary state is a saddle point. When $\pi'(n^*) > 0$ and α are sufficiently small, both eigenvalues are positive and the stationary state is an unstable node. In that case the roots are complex and involve cyclical fluctuations. Finally, since there are three stationary states and the second one is an unstable node, the industry evolution depends on the initial number of firms and its closeness to a specific steady state.

Folster and Trofimov estimate the entry equation (5.25a) on the basis of panel data from the survey data of 360 Swedish industrial firms over 1988–90 and found empirical support for the notion that positive entry is induced by the condition $v > z$ and that the profit function v is S-shaped, which implies three phases of the three stationary states.

Several lines of generalization of this dynamic model are possible. First of all, entry may be viewed in terms of industry output y rather than n, so that the entry dynamic becomes $\dot{y} = \alpha(v - z)$, that is, if profit rises $v > z$, then the output growth is positive in the long run. This implies optimal capacity expansion. For firm j this becomes $\dot{y}_j = \alpha_j(v_j - z_j)$. Let C_j be the net cost of entry $(z - v)$ depending on output y_j and D be total industry demand. Industry equilibrium requires an optimal number n_j

of firms of type j determined by minimization of total industry costs

$$\mathrm{TC} = \sum_j n_j C_j(y_j)$$

subject to

$$\sum_j n_j y_j \geq \mathrm{D}$$

On using p as the Lagrange multiplier associated with demand–supply equilibrium, one may easily determine the optimal values $\left(y_j^*, n_j^*\right)$ from the following conditions

$$p_j = mc_j, \quad mc_j = \partial C_j(y_j)/\partial y_j$$
$$mc_j = ac_j, \quad ac_j = C_j(y_j)/y_j.$$

Second, the R&D effort variable may be usefully replaced by a cost function $c(u) = c(u, n)$ depending on own effort and the industry-wide knowledge base proxied by n and the optimal control path $u = u(t)$ determined by maximizing an adjusted profit function as

$$\hat{v} = \int_t^\infty \mathrm{e}^{-r(\tau - t)}[\pi(n, u) - c(n, u)]\,\mathrm{d}t.$$

This type of formulation would specifically allow us to evaluate the impact of spillover R&D. Finally, the profit function $\pi(n, u)$ may be extended to include the impact of risk due to uncertain future demands, for example, π may be replaced by $U(\overline{\pi}, \sigma^2)$, where $\overline{\pi}$ and σ^2 are the mean and variance of profits and $U(\cdot)$ is an indirect utility function.

5.3 Capacity expansion and growth

The concept of core competence in modern organization theory is basically related to the notions of long-run optimal growth through managerial innovations. This growth objective provides a rationale for the firm to pursue strategic decisions for optimal capacity expansion. The stock of knowledge capital embodied through own R&D (u) and the intra-industry R&D (u_S) due to spillover provides for the modern technology-intensive industries a critical source of growth of output (g). The decision to choose an optimal output growth thus amounts to a decision to choose optimal levels of R&D expenditures u and u_S. Baumol (1967) used this growth strategy to formulate a profit maximizing growth equilibrium model of a firm, where the growth rate $g = \dot{y}/y$ is used as the central decision variable instead of the level of output (y). On using net present value (NPV) as the revenue function $P(g)$ and $C(g)$ as the cost of expansion, the objective function is written as $J(g) = P(g) - C(g)$, where

$$P(g) = \sum_{t=0}^{\infty} R_0 \left(\frac{1+g}{1+r}\right)^t = R_0 \frac{1+r}{r-g}, \quad g < r$$

$C(g)$ strictly convex in g.

Here R_0 is the initial revenue and r is the firm's cost of capital determined by the outside capital market and the condition $g < r$ is assumed for convergence.

This model fails to incorporate however the role of research and knowledge capital either as own research u or the spillover research u_S absorbed by the firm. Second, it ignores the entry–exit dynamics in industry evolution. The *net* cost of entry e may be viewed as the difference $z - v$ of opportunity cost of entry (or cost of increasing market share) denoted by z and the market value of the incumbent firm. On incorporating these two generalizations the growth efficiency model of a leading firm may be formulated as

$$\underset{u,\theta}{\text{Max}}\, J = P(g) - C(g), \quad g = g(u, \theta u_S, e) \tag{5.27}$$

Here θ denotes the proportion of intra-industry spillover R&D that the incumbent firm can absorb. Clearly u and θ are interdependent, since increasing own R&D increases the firm's capability to absorb spillover research and develop its own R&D. Denoting partial derivatives by subscripts, for example, $P_g = \partial P(g)/\partial g$ it is assumed that

$$g(u) > 0, \quad g_\theta > 0 \quad \text{and} \quad g_e < 0.$$

The optimality conditions following from (5.27) are

$$(P_g - C_g)g_u = 0 \quad \text{and} \quad (P_g - C_g)g_\theta = 0$$

and these are sufficient if the Hessian matrix H is negative definite, where

$$H = \begin{pmatrix} J_{uu} & J_{u\theta} \\ J_{\theta u} & J_{\theta\theta} \end{pmatrix}.$$

If θ and u are independent, then $J_{\theta u} = J_{u\theta} = 0$ and J_{uu} is negative if $P_{gu} > 0$ with $P_{gu} < C_{gu}$.. Likewise, $J_{\theta\theta}$ is negative if $P_{g\theta} > 0$ with $P_{g\theta} < C_{g\theta}$.

One basic limitation of this approach is that it is essentially static since there is no path dependence. The assumption of g less than r reduces the objective function to a static form. However in many situations of industry evolution output grows at a rate higher than the interest rate. Hence it is preferable to develop an alternative framework of growth efficiency, where firms are assumed to maximize a discounted sum of dividends subject to a cost of expansion. Let $K(t)$ denote the composite knowledge capital including own R&D and a proportion of spillover industry capital such that

$$\dot{K} = I - \delta K, \quad \dot{K} = \mathrm{d}K/\mathrm{d}t \tag{5.28a}$$

I being gross investment with δK as total depreciation. Dividend is $d = (1-\alpha)\pi(t)$, where the amount of profit $\alpha\pi(t)$ is used for investment for capacity expansion so that the expansion cost $C(I)$ equals $\alpha\pi(t)$, that is,

$$\alpha\pi(t) = C(I) \tag{5.28b}$$

Also profit due to expansion must satisfy a lower bound $\pi_0 > 0$, that is,

$$\pi(K) \geq \pi_0 \tag{5.28c}$$

The firm's objective is to maximize the objective function

$$\text{Max} \quad J = \int_0^{\infty} \mathrm{e}^{-rt} W(d)\,\mathrm{d}t \tag{5.28d}$$

subject to (5.28a) through (5.28c)

This model is more general in several ways, for example, dividend growth is implicit in the integrand $W(d)$ and the optimal time-path of the R&D capital $K(t)$ is solved from this model, where the investment (I) in knowledge capital is the control variable. On using the current value Hamiltonian

$$H = \mathrm{e}^{-rt}[W(d) + \mu(I - \delta K) + \lambda(\alpha\pi(t) - C(I)) + \gamma(\pi(t) - \pi_0)]$$

the necessary conditions yield the following:

$$\dot{\mu} = (r+\delta)\mu - \gamma\,\delta\pi/\delta K, \quad K = I - \delta K$$

$$\delta W/\delta d = \lambda$$

For concave $W(d)$ these conditions are also sufficient, if we include the transversality condition $\lim_{t\to\infty} e^{-rt}\mu(t) = 0$. The characteristic equation yields the two eigenvalues

$$\theta = (1/2)[r \pm (r^2 + 4\delta(r+\delta))^{1/2}] \tag{5.29}$$

if we linearize $\delta\pi(t)/\delta K$ as $\tau_0 + \tau K(t)$.

It is clear from (5.29) that the two eigenvalues θ_1, θ_2 are real and opposite in sign and hence there is a saddle point equilibrium for the steady state as in the model of Folster and Trofimov. In the steady state we get

$$\mu^* = (r+\delta)^{-1}(\gamma(\tau_0 + \tau K^*)) \tag{5.30}$$

$$K^* = I^*/\delta$$

This type of model has been generalized by Sengupta (1972) to multi-product firms, where balanced growth solutions may be specified as the optimal expansion path.

Note that this type of dynamic model cannot generate any cyclical movements near the steady state, since the eigenvalues are all real. Thus the equilibrium path defined by the adjoint equation converges to a long run equilibrium defined by (5.30). Also the control variable $I(t)$ may be separated into investment for own R&D and the external spillovert proportion appropriated by the incumbent firm. The entry dynamics in this framework would be generated by new firms entering the industry, wherever the incumbent firm fails to remain on the dynamic optimal trajectory of R&D investment. Since optimal R&D investment helps to reduce the average long run cost c_i of the incumbent firm i, one could easily estimate from empirical panel data the impact of I^*:

$$\dot{s}_i = a_i s_i(\bar{c} - c_i), \quad a_i > 0 \tag{5.31}$$

where s_i is the market share of firm i and $\bar{c}$ is the industry average cost. If c_i falls when I^* rises, the market share of i increases. If firm i is an outsider and it is able to reduce c_i below $\bar{c}$, then there is more entry. In the opposite case, if c_i exceeds $\bar{c}$ then there is cost pressure for the incumbent firm to exit.

Concluding remarks

Innovation in different forms affects the selection and industry evolution in two important ways. One is that it helps to reduce average cost relative to the industry level thus improving the long-run cost frontier. Second, research and development can yield sufficiently strong externalities that can exhibit aggregate IRS industry level R&D. The latter may provide strong incentives to form implicit collusion and/or mergers. These aspects are important for designing optimal competition policies for any industry where R&D investment is very high, for example, pharmaceuticals or consumer electronics.

References

Baumol, W. (1967): *Business Behavior, Value and Growth.* Harcourt, Brace and World, New York.

Cohen, W.M. and Levinthal, D.A. (1989): Innovation and learning: The two faces of R&D. *Economic Journal* 99, 569–96.

Denny, M., Fuss, M. and Waverman, L. (1981): The measurement and interpretation of total factor productivity in regulated industries with an application to Canadian telecommunications. In: Cowing, T. and Stevenson, R.E. (eds), *Productivity Measurement in Regulated Industries.* Academic Press, New York.

Folster, S. and Trofimov, G. (1997): Industry evolution and R&D externalities. *Journal of Economic Dynamics and Control* 21, 1727–46.

Jovanovic, B. (1997): Learning and growth. In: Kreps, D.M. and Wallis, K.F. (eds), *Advances in Economics and Econometrics.* Cambridge University Press, New York.

Levin, R.C., Nelson, R.R. and Winter, S.G. (1987): Appropriating the returns from industrial R&D, Brookings Papers on Economic Activity. Washington, DC.

Lucas, R.E. (1993): Making a miracle. *Econometrica* 61, 251–72.

Prahalad, C.K. and Hamel, G. (1994): *Competing for the Future.* Harvard Business School Press, Cambridge.

Sengupta, J.K. (1972): Balanced growth path for an expanding multiproduct firm. *International Economic Review* 13, 553–67.

Sengupta, J.K. (2003): *New Efficiency Theory.* Springer-Verlag, Berlin.

Sengupta, J.K. (2004): Dynamic efficiency with learning by doing. *International Review of Applied Economics* 47, 95–101.

6
Competition and Innovation Efficiency

The complexity of the innovations process and the diverse forms it can take raise two critical issues for the modern technology-intensive firms. It raises the development costs both internally and externally, internally because the firms may not have all the necessary assets or strongholds and externally because the firms may lack the opportunities for appropriating the intra-industry R&D knowledge capital. Secondly the technology strategy and its economic aspects, for example, how to choose between fixed and flexible technology, have strongly influenced the risk in organizational environment. To reduce these twin costs, that is, the cost of development and the cost of risk and uncertainty, the incentives for collaboration and/or networking in R&D have increased significantly. The theorists have argued that if competitors are not allowed to collaborate in R&D, then they may not invest enough because of the free rider problem. Collaboration and/or cooperation in R&D may also lead to collaboration on the product and marketing side, thus affecting the entry–exit dynamics, for example, it may prevent potential entry by outsiders, because of the scale effects of pooled knowledge capital. Policymakers at the state level have increasingly been aware of these circumstances and recognized the fact that most firms do not have all the necessary assets, both physical and human for developing latest technologies and staying on the edge of the dynamic cost frontier based on innovation. By now policymakers in many countries have explicitly recognized the firms' needs to form collaborative activities in R&D innovation. In Europe, Japan and the US firms are now allowed to form R&D cooperatives in some form . The fast growing countries of Southeast Asia called the four tigers (S. Korea, Singapore, Taiwan and Hong Kong (China)) have explicitly encouraged the formation of such cooperation and been provided significant incentives by the state in their fiscal and monetary policies. The latter has made a significant contribution to the phenomenal growth of exports of technology-intensive products from these countries in the last two decades.

Our objective here is threefold. One is to discuss the models of cooperative and collusive behavior in R&D expenditures and their implications for efficiency gains due to cooperation. We discuss in particular the seminal contribution of d'Aspremont and Jacquemin (1988), which shows that cooperative R&D levels exceed the noncooperative levels whenever the extent of spillovers is relatively large, that is, above 50%. Second, we discuss the role of an optimal competition policy by the state, when partial collusion in R&D technologies may be allowed but no collusion on the product market. Finally, we discuss the economies to be gained by a technology strategy for cooperative R&D. In this context we analyze the growth and efficiency of the computer industry over the period 1985–2000 and the critical role played by R&D inputs.

6.1 Collusion and cooperation in R&D

Modern firms in the information technology sector today have several economic incentives to cooperate and combine R&D efforts. First of all, the technology of the new innovative process is becoming increasingly complex and the development cost is very large. Second, not all results of process innovations in new technology can be appropriated by the inventor, since the competitors will copy the invention and thus "free ride". Third, collusion and cooperation in the R&D phase may help the inventor firms to internalize a large proportion of the "spillovers" and thereby reduce unit costs and gain larger market shares. By now the policymakers in most developed countries have recognized this need by firms to cooperate in R&D and related innovations. For example, the European Commission (EC) allowed in March 1985 a 13-year block exemption under Article 85(3) of the Treaty of Rome to all firms forming joint ventures in R&D.

Philips (1995) has argued that joint R&D ventures have several economic advantages. They are more flexible forms of organizations than mergers. Collaborative research also allows the firms to spread risks and sunk costs and to exploit large scale economies particularly when there exist significant indivisibilities. There also exist disadvantages. Once the competitors are allowed to cooperate/collude in the process stage, they also get the incentive to collude in the product market, which is not permitted by the anti-trust laws. Also the competitors who form the flexible form of collaborative R&D rather than a merger which is a more rigid form of organizational structure, have always the incentive to break away. This frequently happens in a cartel, where implicit cheating is very often the rule rather than the exception.

d'Aspremont and Jacquemin (1988), henceforth referred to as A&J are perhaps the first to analyze the collaborative R&D situations and compare them with noncooperative R&D levels. Their major conclusion is that optimal cooperative R&D levels exceed those of noncooperative R&D situations whenever technological spillovers are relatively large (i.e., above 50%), while the opposite holds for small spillovers below 50%. Hinloopen (2003) has recently extended this model in two directions. One is the case when the R&D spillovers are relatively small but still a cooperative agreement is likely to raise effective R&D efforts, if the agreement induces a sufficient increase in technological spillover. Secondly, he models the R&D spillover in terms of R&D inputs rather than outputs. A&J model considers only R&D outputs as spillover effects after the R&D process is finished. Modeling spillover in terms of inputs appears to be more realistic, since the research exchange is mainly done through research efforts disseminated through networking channels. We discuss first the R&D efficiency gains through cooperation in research activity through Hinloopen's model and then discuss the A&J model and its generalizations.

Assume two duopolist firms facing market demand

$$p = a - bY, \quad Y = y_1 + y_2 \tag{6.1a}$$

where y_i is output of firm i, which invests r_i in R&D and receives a share $\beta r_j (i \neq j), 0 < \beta < 1$ of spillover, that is,

$$R_i = r_i + \beta r_j, \quad i \neq j, \quad i, j = 1, 2 \tag{6.1b}$$

Its cost function is assumed as

$$C_i = A - f(R_i), f(R_i) = (R_i/\gamma)^{1/2}$$

where $f(R_i)$ is the production function assumed to be of the form above showing that $f(\cdot)$ is increasing in R_i at a decreasing rate with $f(0) = 0$. On using the profit function

$$\pi_i = py_i - C_i y_i - r_i \tag{6.1c}$$

If each firm behaves noncooperatively as Cournot players in the R&D stage, one obtains the optimal noncooperative level of R&D as R^{NC}, where the output of R&D is

$$f(R^{NC}) = [(a - A)(2 - \beta)]/[9b\gamma - (2 - \beta)]$$
$$= [(1 + \beta)R^{NC}/\gamma]^{1/2} \qquad (6.1d)$$

Similarly the cooperative optimal solution R^C is obtained by maximizing joint profit $\pi(r_i, r_j) = \pi_1 + \pi_2$ with

$$f(R^C) = [(a - A)(1 + \theta)]/[9b\gamma - (1 + \theta)]$$
$$= [(1 + \theta)R^C/\gamma]^{1/2} \qquad (6.1e)$$

Two implications easily follow: (i) if $0 \leq \beta < 1/2$, then $R^C < R^{NC}$, (ii) if $1 - \bar{\beta} < \beta < \bar{\beta}$, then $R^C > R^{NC}$, where $0 \leq \beta < \bar{\beta}$ and $0 < \bar{\beta} < 1$ if and only if $\theta - \beta \geq \max\{0, 1 - 2\beta\}$. Thus a substantial increase in pre-cooperative R&D spillover is necessary for the combined joint profit externality to be strong enough to induce the R&D cooperative to do more R&D than would a competitive R&D market. On the other side if $\bar{\beta}$ is less than 50% the post-cooperative spillover can never be such that R&D efforts exceed those of a competitive R&D market. Furthermore, one obtains the optimal cooperative profits π^C as

$$\pi^C = [\gamma(a - A)^2]/[9b\gamma - (1 + \theta)]$$

which implies that firms always attempt to increase their spillover as much as possible if they form an R&D cooperative, since $\partial\pi^C/\partial\theta > 0$. This result reinforces the proposition that the free flow of information between competitive firms will increase if they enter into a cooperative agreement.

A&J consider a quadratic cost function $C_i = (A - r_i - \beta r_j)y_i$ with $0 < A < a$, $0 < \beta < 1$, $r_i + \beta r_j \leq A$ and $Y = y_1 + y_2 \leq a/b$. The optimal strategy of each firm is how to choose R&D level (r_j) in the first stage and a subsequent level of output (y_i) in the second stage. Three different games are considered with two stages in each. The first stage determines R&D and the second the output. The first one assumes that firms act as Cournot–Nash players in both output and R&D. When firms choose R&D levels, profits are

$$\pi_i^* = (9b)^{-1}[(a - A) + (2 - \beta)r_i + (2\beta - 1)r_j]^2 - (\gamma/2)r_i^2;$$
$$i \neq j; \quad i, j = 1, 2 \qquad (6.2)$$

Maximizing π_i^* with respect to each r_i yields the optimal values

$$r_i^* = [4.5b\gamma - (2-\beta)(1+\beta)]^{-1}[(a-A)(2-\beta)] \quad i = 1,2$$
$$Y^* = y_1^* + y_2^* = (3b)^{-1}[2(a-A) + 2(\beta+1)r_i^*]$$
$$= [(3b)(4.5b\gamma - (2-\beta)(1+\beta))]^{-1}[9(a-A)b\gamma] \qquad (6.3)$$

The second model assumes cooperative R&D in the first stage and noncooperative output in the second stage. On maximizing joint profits in the first stage the optimal symmetric solution $\hat{r}_1 = \hat{r}_2 = \hat{r}$ is obtained as

$$\hat{r} = [4.5b\gamma - (\beta+1)^2]^{-1}[(\beta+1)(a-A)]$$
$$\hat{Y} = [3b(4.5b\gamma - (1+\beta)^2)]^{-1}[9b\gamma(a-A)] \qquad (6.4)$$

Clearly for large spillovers, that is, $\beta > 0.5$, we get $\hat{r} > r^* = r_1^* = r_2^*$ indicating that for large spillovers R&D level increases when firms cooperate in R&D in the first stage. Also $\hat{Y} > Y^*$, that is, total output level is also higher.

The third case assumes cooperation or collusion in both stages. The optimal solutions are

$$\tilde{r} = [4b\gamma - (1+\beta)^2]^{-1}[(a-A)(1+\beta)]$$
$$\tilde{Y} = (2b)^{-1}[(a-A) + (1+\beta)\tilde{r}]$$
$$= [2b(4b\gamma - (1+\beta)^2)^{-1}[4b\gamma(a-A)] \qquad (6.5)$$

It is clear that for a given level of R&D, the total collusive output is smaller than the noncooperative one, but not necessarily so when the optimal R&D is incorporated. Also the collusive level of R&D varies with the level of β, for example, for large spillovers it is higher than in the fully noncooperative Cournot–Nash equilibrium in both stages.

A&J also consider a socially optimal solution obtained by maximizing social welfare $W(Y)$ defined as the sum of consumers' surplus and the producer surplus conditional on $r = r_1 = r_2$.

$$r^{**} = [2b\gamma - (\beta+1)^2]^{-1}[(a-A)(1+\beta)]$$
$$Y^{**} = (1/b)[(a-A) + (1+\beta)r^{**}] \qquad (6.6)$$

Clearly $r^{**} > r^*$, since

$$[2b\gamma - (1+\beta)^2]^{-1}(1+\beta) > [4.5b\gamma - (2-\beta)(1+\beta)]^{-1}(2-\beta)$$

and also $\quad Y^{**} > Y^* \qquad (6.7)$

Furthermore $r^{**} > \tilde{r} > \hat{r}$ and $Y^{**} > \hat{Y} > \tilde{Y}$.

Philips (1995) has discussed the important policy implications of these results. For example, the anti-trust authorities should encourage the formation of joint ventures in R&D implying a sharing of all information through making γ equal to one but without allowing collusion in the product market. When cooperation is possible in both stages, optimal R&D levels are even greater, that is, $\tilde{r}_i > \hat{r}_i$. Collusion in the product market allows firms to capture more of the surplus created by R&D and induces more R&D investments.

Kamien, Muller and Zang (1992) have generalized the A&J approach by considering differentiated products, more than two firms and Bertrand games. Although this leads to a large number of games, the general results remain the same, that is, when joint profit maximization is combined with full cooperation, that is, $\gamma = 1$, optimal R&D investments are even higher than $\hat{r}_i$ and the total output is higher than $\hat{Y}$ thus implying lower prices.

Sometimes in the product market tacit collusion may become an equilibrium strategy. In a market system the firms invest in new technology and new product development because of the profit they expect to earn after discovery and development. Thus a firm's incentive to invest in R&D depends on the difference between pre-innovation profit and the profit it expects to earn after innovation. Competition policy by the state typically prohibits explicit collusion, which may nonetheless occur. However these profit-maximizing firms may alter their behavior to take expected anti-trust penalties into account. Thus the colluding firms may raise the price above the noncooperative equilibrium level but deliberately hold price below the joint profit-maximization level to reduce the probability of attracting attention of anti-trust agencies.

6.2 Entry dynamics and innovation

What determines the probability of entry and of exit, that is, the actual net entry rate into a new market or industry developed by Schumpeterian innovation? Gort and Konakayama (1982) have developed a stochastic model of entry and empirically estimated it. Let $\pi(z)$ be an objective function, for example, profits depending on a set of variables z, such that $z(1)$ is the value of z if the firm enters and $z(0)$ if it does not. The firm chooses entry if $\pi[z(1)] > \pi[z(0)]$. The probability of entry P is then given by

$$P = \text{prob}[\pi(z(1)) > \pi(z(0))] \tag{6.8a}$$

For simplicity of estimation they specify $[\pi(z(1)) - \pi(z(0))]$ as a linear function of the explanatory variables X, that is,

$$\pi(z(1)) - \pi(z(0)) = X'\beta$$

where X is a vector of observable variables, prime denoting transpose. Let $E = G - D$ denote the difference of expected number of gross entrants (G) and exists (D) in year t. Then one sets up the following estimating function for net entry:

$$E_t = P_t(\overline{N}_{t-1} - n_{t-1}) - Q_t n_{t-1} + u_t \tag{6.8b}$$

where P_t and Q_t are respectively the probabilities of entry and of exit, $\overline{N}_{t-1}$ is the population of potential entrants, n_{t-1} is the number of existing firms in the market at the end of time $t - 1$ and u_t is a stochastic disturbance term. For estimation purposes we assume the us to be independently normally distributed with mean zero. The sample data on product innovations was selected from a set of 46 product histories developed in a research survey by Gort and Kleeper (1982), annual data on patents were obtained from the US Patent Office and the variable $\overline{N}$ was based on data from the Census of Manufactures for the years 1947, 1954, 1958, 1963, 1967 and 1972. Table 6.1 shows the elasticities of estimated gross entry and estimated gross exit with respect to the following explanatory variables: $X_1 = (n/\overline{N})_{t-1}, X_2 =$ growth rate in output per firm $= \Delta(y/n)_t/(y/n)_{t-1}$, $X_3 = \Sigma y_{t-1}/\Sigma y_T$, $X_4 =$ number of patents issued and $X_5 =$ number of major innovations as classified in the research survey.

Table 6.1 shows several interesting findings. First, a 1% change in major innovations shows a 3.65% change in net entry. Second, the innovations embodied in the patent rates (X_4) have the effect of increasing entry,

Table 6.1 Elasticities of gross entry and exit with respect to X_i

	Gross entry	**Gross exit**	**Net entry**
X_1	13.60	—	9.66
X_2	1.17	0.41	0.63
X_3	3.05	0.89	2.01
X_4	4.05	0.51	2.90
X_5	4.22	−0.14	3.65

that is, 4.05 for estimated gross entry and 2.90 for net entry. The growth rate in output per firm (X_2) has the effect of raising gross entry by 1.17% and net entry by 0.63%. The major force that explains exit from the population of existing firms is the relative advantage of firms with more intangible capital like R&D and learning experience, which are proxied here by the patent rate and the accumulated stock of experience of producers in the industry.

Reduction in dynamic adjustment cost is here proxied by the variable X_2, since a barrier to competition may arise from the superior efficiency of existing firms, in which case their low prices can deter rivals. Demsetz (1982) has argued that this type of barrier may involve predatory pricing behavior, that is, reducing price now, presumably below cost in the hope of giving rise to a future stream of higher prices.

Innovation efficiency through R&D expenditure and shared cooperative investment helps to deter potential entry through implicit barriers to entry. Bain (1956) has given three main reasons for barriers to entry: absolute cost advantages of the incumbent firm, economies of scale and product differentiation advantages. von Weizsacker (1980) and more recently Geroski (2003) have used the principle of absolute cost advantages and scale efficiency as the basis of superior efficiency. If an incumbent firm is more efficient than a potential entrant, then the incumbent firm will continue earning an efficiency rent and deter new entry, unless entrants can be found who are just as efficient as the incumbent firms with superior efficiency. The following entry dynamics capture this characteristic:

$$\begin{aligned} E_{it} &= \alpha(n^*_{it} - n_{i,t-1}), \quad 0 \leq \alpha \leq 1 \\ n^*_{it} &= \mathrm{TC}(y^e_{it}) \end{aligned} \tag{6.9}$$

where E_{it} is net entry in industry or product market in industry at time t, n_{it} the actual number of producers in the industry, n^*_{it} the anticipated cost-minimizing number of producers in the industry and $\mathrm{TC}(\cdot)$ relates the anticipated cost-minimizing number of producers to the anticipated equilibrium level of output y^e_{it} at time t. von Weizsacker (1980) has considered models of sequential innovations in technology-intensive modern industries, where the expected price falls exponentially through time $p^*_t = p_0 \, exp(-\gamma t)$, where the progressiveness of the industry is measured by γ, the rate of decline of prices. This parameter is mainly determined by the rate of technological progress induced by the innovations. Thus the lower the expected price p^*_t, the lower is the net entry rate. Note however that the parameter γ is related to the market size effect. For example if $y = ap^{-\varepsilon}$ is the demand function, then the effect of the price elasticity of demand on the R&D externalities can be explained by the market size effect of ε. If the price elasticity is higher, the market grows faster through time with declining prices. A rapidly growing market thus is an indicator of high externalities or spillover of innovation. Thus γ is an endogenous parameter depending on the speed with which innovations are implemented. For example, a higher price elasticity of demand by the market size effect will induce a greater speed of sequential innovations and hence a higher value of γ.

The basis for competition in modern progressive industries has changed dramatically in the past two decades, thanks to the developments in software technology and sophisticated technical progress in telecommunications, microelectronics and the computer industry. Modern competition starts basically at the product design stage. The survival of particular firms at this stage depends on the market viability of their design. Once the dominant design emerges, competition focuses on price. A process of consumer learning on the demand side of the market reinforces the short-run forces to reduce the costs and prices. Geroski (2003) has characterized the evolution of new product markets in terms of two types of technology: *sustaining technologies* and *disruptive technologies*. The former enhances competence by extending an existing technology, for example, the endless increase in computer memory observed in recent decades. Disruptive technologies tend to bring a new dominant product or process design, usually from outside the market. This is very similar to the destructive aspect of Schumpeterian process of creative destruction.

One important feature of the entry–exit dynamics is its path dependence on the technological or product diffusion path of an innovation or a sequence of innovations. Strong empirical evidence exists to support the view of an S-shaped diffusion path of an innovation, for example, Griliches (1957) and more recently Gort and Kleeper (1982) have demonstrated this very clearly. Jovanovic and Lach (1989) have shown how this S-shaped diffusion path arises naturally in a competitive framework, where homogeneous agents face the prospect of learning-by-doing in Arrow's sense. They define the act of adopting a new technology and/or starting production of a new product as *entry*, and that of ceasing production as *exit*. The S-shaped diffusion path generates staggered entry and exit. Thus early entry has the advantage of higher revenues per unit of output but late entry has the benefit of learning from the experience of earlier entrants, thereby lowering unit production costs. Their model measures cumulative gross investment by n_t, the number of machines (designs) installed before time t. Each machine (design) has a fixed installation cost $k(n_t)$ and a variable unit cost $c(n_t)$ of production with declining marginal costs and a fixed capacity constraint. As entry proceeds and the product price p_t declines, then some machines (designs) are withdrawn as soon as p_t equals its marginal cost, that is, vintage s is withdrawn as soon as $p_t \leq c(n_s)$. Denoting by x_t the cumulative exit up to time t, one obtains

$$x_t = c^{-1}(\min[p_t, c(0)]) \equiv x(p_t) \tag{6.10a}$$

They assume market clearing so that total industry supply at t is $n_t - c_t$ which equals market demand $D(p)$. Let $P = P(n_t)$ be the market equilibrium price, then the exit rate $\dot{x} = dx/dt$ may be computed from (6.10a) as

$$\dot{x} = x'P'\dot{n} \tag{6.10b}$$

where prime denotes differentiation and

$$P' = 1/(D' + x') \text{ so that } \dot{x} = [x'/(D' + x')]\dot{n} < \dot{n}.$$

Net entry defined by $\dot{n} - \dot{x}$ is always positive, since

$$\dot{n} - \dot{x} = [D'/(D' + x')]\dot{n} > 0$$

so that the number of firms and hence the total industry output $n - x = n - x[P(n)]$ grows as long as $\dot{n}$ is positive.

One can also determine total gross entry N in equilibrium. This number is such that further entry would yield negative discounted profits. Hence N satisfies

$$r^{-1}[P(N) - c(N)] - k(N) = 0 \tag{6.10c}$$

where r is the discount rate used for the discounted profit function, the instantaneous profit being $\pi = P(n) - c(n) - rk(n)$. Assuming that a unique solution N exists for (6.10c), they have derived an important result as follows: A necessary and sufficient condition for a positive gross exit is that the average cost of the last entrant is less than the marginal cost of the first entrant, that is,

$$c(0) > c(N) + rk(N)$$

This follows from (6.10c) since $p(N) < c(0)$.

Consider a simple example showing the time-path of the diffusion curve.

Let $k(n) = An^{-\alpha}, \mathrm{D}^{-1}(n) = Bn^{-\beta}, 0 < \alpha < \beta < 1$. Then from (6.10c) it follows

$$N = (B/rA)^{1/\lambda\alpha}, \quad \lambda = (\beta - \alpha)/\alpha > 0.$$

When the initial condition is $n_0 = 0$ one obtains

$$n_t = N(1 - e^{-\lambda rt})^{1/\lambda\alpha}$$

Note that the competitive equilibrium rate of entry is too low from the social welfare viewpoint. This is due to the spillover effect flowing to the future entrants. A social planner therefore has to build a cooperative mechanism for sharing R&D investments.

6.3 Efficiency and growth in the computer industry

If superior efficiency through dynamic cost minimization and price reductions can be used as a potential entry-preventing strategy and competitors are allowed to form cooperation in the R&D stage but not in the product market stage, then the computer industry today would provide a fitting example for studying the impact of innovation efficiency through R&D on the competitive growth of leading firms. Here entry (exit) has to be defined more broadly as the market share of a firm above (below) the industry average. Then one can assess the role of R&D expenditure on the entry–exit dynamics of firms.

The computer industry has undergone rapid changes in technology over the last three decades. Norsworthy and Jang (1992) are probably the first to estimate the productivity growth in the computer industry comprising mainframe, mini and microcomputers over three subperiods 1959–67, 1967–75 and 1975–81. In a simplified form they estimated the translog cost function as

$$\ln \mathrm{TC} = b_0 + b_1 \ln y(t) + b_2 T \tag{6.11}$$

where $y(t)$ is output and T is used as a time trend variable proxying the state of technology. The inverse of the parameter b_1 can be interpreted here as the degree of returns to scale. A negative value of b_2 indicating a pure shift of the cost function is then a measure of Hicks-neutral technical progress.

On differentiating (6.11) one obtains

$$\Delta \text{TC}/\text{TC} = b_1(\Delta y/y(t)) + b_2$$

The estimate for b_2 turns out to be (−0.0369) for the whole period 1959–81, whereas $\hat{b}_1 = 0.3477$ and both are significant at 10% level. This suggests that the long-run average rate of technical change (or progress) is 3.69% per year in the US computer industry over 23 years 1959–81 and the degree of returns to scale (S) is $1/\hat{b}_1 = 2.876$. Both are substantial. The technical change for the three subperiods 1959–67, 1967–75 and 1975–81 were −0.037, −0.051 and −0.041 suggesting no definite trend, although when production workers ($b_{2\text{L}}$) and nonproduction workers ($b_{2\text{N}}$) are used as explanatory variables, the degree of production-worker saving decreases through time from $\hat{b}_{2\text{L}} = -0.011$ in 1959–67 to $\hat{b}_{2\text{L}} = -0.004$ in 1975–81.

The technological developments in the US computer industry since 1981 have followed two distinct patterns in the new information age. One is the important role of R&D investment in both human capital and shared investment network. The theory predicts that technical uncertainties in R&D investment outcomes can be hedged considerably by pursuing multiple conceptual approaches in parallel. Parallel as opposed to serial exploration of alternative concepts reduces the expected time to successful project completion but raises R&D costs but this can be considerably reduced by sharing and networking. The US computer industry has played a dynamic role in this connection due to the stiff challenges from abroad and rapid growth of international trade. The second important trend in the US computer industry is the pattern of product cycle and its changes over time in recent years. Some researchers have argued that in the computer industry the survival of the fittest depends on the need for a company (firm) to bring on line a continuous stream of new products. Each product goes through a typical life cycle of R&D, market introduction, maturation and eventual obsolescence. Two trends are important in this evolution: the market initially expands (this is sometimes called economies of scale in demand) and competition becomes more intense. To supply the expanding market, scale economies and learning curve effects are exploited along with technical progress and this is accompanied by falling prices. Thus many early niche markets in the computer industry such as the markets for laptops or palm PCs have now grown into mass markets. As the manufacturers scramble to shorten the time to market, the R&D and commercialization phase of each cycle becomes shorter. As the products gradually penetrate an ever expanding market, the upswing and maturation phases become longer and longer thus generating accelerated growth.

Costs of manufacturing exhibit significant variations over the product cycle. These costs fall due to learning curve and a significant degree of miniaturization.

In the computer industry no company of course stays with a single life cycle due to a rapidly changing technology pattern. The managerial challenge is to bring an optional stream of new "vintages" of its products and as each company does so, the process of technological evolution continues forward in the industry.

There exist however some stylized facts in the recent trend in the computer industry. One is the speed of technical change and the second is the growth in demand.

The impact of this technical change can be modeled in terms of the change in market share s_i of firm i, which may be used a proxy for entry or exit. The entry dynamics then take the form

$$\dot{s}_i = \lambda s_i(\bar{c} - c_i) \tag{6.12a}$$

where dot denotes time derivative, c_i is the minimal average cost of firm group i and $\bar{c}$ is the industry level average cost. Here λ determines the speed at which firm's market penetration adjusts to differences between c_i and $\bar{c}$. Thus a high value of λ indicates an industry with a strong competitive adjustment. This may intensify the impact of barrier to entry on potential entry.

One may view $c_i = c(\mathrm{I}_i) = -h\mathrm{I}_i(t)$ as a function of innovations in R&D denoted by $I_i(t)$ with its parameter $(-h)$ reflecting its cost reducing aspect. The industry average cost $\bar{c}$ represents the mean fitness of an average firm. The source of potential dominance of firm is measured here by the gap $(\bar{c} - c_i)$, where $c_i < \bar{c}$. One of the stylized facts of the entry–exit dynamics noted by the survey results by Lansbury and Mayes (1996) is that a growing competitive industry involves not just the expansion of existing firms but also new entrants who challenge the incumbents often with new innovations and thus the process of creative destruction of the old processes and products begins. This has been called the *churning process*, one consequence of which is that the higher the heterogeneity in the industry, the larger the exit rate. Following Fisherian selection dynamics this may be characterized as

$$d\bar{c}/dt = \dot{\bar{c}} = -\alpha\sigma^2(\mathrm{t}), \quad \alpha > 0 \tag{6.12b}$$

where the variance $\sigma^2(t)$ may be used as a measure of heterogeneity in firm sizes in the industry or the product. On combining (6.12a) with (6.12b) we obtain the second order dynamics

$$\ddot{s}_i - a_1\dot{s}_i + a_2 s_i = 0, \quad a_1 = \lambda k_i, \quad k_i = \bar{c} - c_i, \quad a_2 = \lambda(\alpha\sigma^2 - h\text{I}_i(t)).$$

Conditional on a given level of k_i, the characteristic equation of this system can be easily computed with two eigenvalues μ_1, μ_2 as

$$\mu_1, \mu_2 = (1/2)[a_1 \pm (a_1^2 - 4a_2)^{1/2}].$$

Clearly is $h\text{I}_i(t) > \alpha\sigma^2$ then $a_2 < 0$ and hence the two eigenvalues are real but opposite in sign. Likewise for the case $a_2 > 0$ but $4a_2 < a_1^2$. Hence the steady state equilibrium is a saddle point. In case $a_1^2 < 4a_2$ we have cyclical fluctuations. Higher σ^2 tends to increase a_2 and hence may increase fluctuations if the roots are complex. In recent times the computer industry has gone through both upswings and downswings, the latter being reflected in the exit rates of many small firms. Thus one can identify three phases in this industry: a rising phase where leading firms grow faster, a falling phase when the losing firms decline or exit and cyclical phase, when entry and exit occur at the same time. A higher rate of heterogeneity in the form of larger $\sigma^2(t)$ have intensified the process of exit.

A second stylized fact in the industry is the pattern of growth in demand and also the changes in the price elasticity of demand due to globalization of the market and international spread of business.

The computer industry in the United States is one the top of the fastest growing sectors in the US economy over the 16-year period 1985–2000. The average sales growth of all the companies in the SIC codes 3570 and 3571 is about 12.8% per year on the average for the period 1985–94, and it is slightly higher (13.1%) for 1995–2000. Some companies like Dell and Silicon Graphics grew much faster over the whole period 1985–2000, for example, an average of 22.4% per year.

Demand growth involved intense market competition followed by increasing technological diversity with greater utilization of economies of scope and learning by doing. All these resulted in increasing cost efficiency and falling prices.

It is useful to separate the demand growth and rapid technological evolution for analytical purposes. Although demand growth has played a key role in technological evolution, the latter has advanced on its own due to R&D investment in both hardware and software. The dynamic role of R&D investment in the companies on the leading age of competition had dramatically improved the performance of computers and their diverse range of applications in almost all the manufacturing industries. Technology in new hardware improved due to the development of new software and the latter developed at a rapid rate due to demand from domestic and international users.

Three aspects of the evolution of demand are important in the growth of the computer sector. One is the increase in volume of demand due to globalization of trade. The expansion of international trade in both hardware and software markets has been spearheaded by the rapid advance of software development by the subsidiaries of leading US companies in Asian countries like Taiwan, Korea, Singapore and India. The second aspect of demand growth is due to the significant economies of scale in demand rather than supply. The elasticity of demand with respect to total industrial output has exceeded 2.91 over the whole period 1985–2000, whereas the income elasticity of demand has been about 1.92. Since the value of a network goes up as the square of the number of users, demand growth has generated further investment in expanding the networks through interlocking and other linkages in the network economy. The third aspect of demand growth is the market size effect. If the price elasticity is higher (e.g., it rose from −1.74 to −2.15 over the period 1985–92 to 1993–2000) the market grows faster through time with declining prices. A rapidly growing market thus reflects high R&D externalities or spillovers.

The demand pattern may be modeled as a first order Markov process. Net sales data are used as a proxy for demand. These data are obtained from Standard and Poor's Compustat files for the period 1985–2000. Denoting y_t as net sales in years t the model is of the form

$$y_t = \alpha + \beta y_{t-1} + \varepsilon_t$$

Since the β coefficient indicates the growth parameter its estimates for some selected firms are given in Table 6.2.

On the average for the industry as a whole the annual rate of growth of demand has increased from 8% to 15%. This has helped the leading firms increase their market share through cost efficiency by utilizing the R&D investments both at home and abroad.

The above sales growth estimates were checked for any serial correlation by DW statistic. If there is no serial correlation, the statistic will be very close to two. For positive (negative) correlation the statistic will approach four (zero). All the ten selected companies reported

Table 6.2 Estimates of demand growth and efficiency

	$\hat{\beta}$	$t_{\hat{\beta}}$	$\overline{R}^2$
1. Dell	1.495	41.181	0.994
2. Compaq	1.276	28.728	0.988
3. HP	1.116	26.664	0.986
4. Sun	1.107	31.810	0.990
5. Toshiba	1.043	9.480	0.899
6. Silicon Graphics	0.994	9.470	0.899
7. Sequent	0.990	9.407	0.897
8. Hitachi	0.718	4.607	0.669
9. Apple	0.699	4.427	0.650
10. Data General	0.721	10.212	0.681

above had DW statistics close to two. Hence the hypothesis of positive or negative serial correlation can be rejected.

The third stylized fact is the significant role played by R&D expenditure. Two methods are adopted to quantify this role. One is the non-parametric method of estimating the production frontier by a sequence of linear programs. This determines two subsets, one efficient (i.e., on the frontier) and the other non-efficient (i.e., below the frontier). Based on these subsets we estimate by regression methods the contribution of R&D inputs. For example Fanchon and Sengupta (2001) used the data set from Standard and Poor's Compustat Data base with SIC codes 3570 or 3571 over the period 1979–1998 comprising 204 firms of which 103 were on the production frontier (i.e., efficient). The five inputs were grouped into three subgroups: cost of goods sold (COGS) including production, selling and administrative expenses, research and development expenditures (R&D) and the capital expenditure (CAP). The regression result is as follows with y as net sales as proxy output:

$$y = \underset{(t=2.56)}{71.855} + \underset{(96.81)}{1.146^{**}}\,\text{COGS} + \underset{(29.11)}{3.591^{**}}\,\text{R\&D} + \underset{(3.53)}{0.283^{*}}\,\text{CAP}$$

$\overline{R}^2$: adjusted $R^2 = 0.9992$

Further when the regressions are run separately for the efficient and the nonefficient firms, the coefficients for R&D are about 10% higher, while the coefficients for COGS and CAP are about the same. When each variable is taken in incremental form and as a two-year average the results are as follows with one and two asterisks denoting significant t-statistics at 5% and 1% respectively.

$$\Delta y = -6.800 + \underset{(t=49.84)}{1.169^{**}} \Delta\text{COGS} + \underset{(13.28)}{2.654^{**}} \Delta\text{R\&D} + \underset{(10.50)}{1.053^{**}} \Delta\text{CAP}$$

$$(\overline{R}^2 = 0.978)$$

$$\bar{y} = \underset{(2.76)}{68.89^{*}} \underset{(100.99)}{1.152^{**}} \overline{\text{COGS}} + \underset{(31.77)}{3.589^{**}} \overline{\text{R\&D}} + \underset{(2.91)}{0.232^{*}} \overline{\text{CAP}}$$

$$(\overline{R}^2 = 0.9993)$$

It is clear that the R&D variable has the highest marginal contribution to output. One has to note however that R&D and capital variables affect production with time lags. Fanchon and Sengupta (2001) have investigated how the choice of lagged variables influences long-run efficiency in a hypercompetitive market, where the rapid change in economic environment invalidates the static assumptions of short-run profit maximization. A two-stage method is adopted for this purpose. In the first stage a regression method is applied to pre-select the explanatory variables and then in the second stage only the statistically significant explanatory variables are used to by the nonparametric convex hull method to estimate the efficiency and the production frontier.

The following steps are followed: (1) identify on economic grounds a set of potential explanatory variables that may affect production efficiency, (2) run a regression model to identify the lag structure of all the firms, (3) identify a set of efficient firms (companies) with an initial nonparametric model termed as a DEA (data envelopment analysis) model, (4) run a regression model twice, once to identify the contribution of each explanatory variable and once using only the efficient firms, and finally (5) check whether the identified set of explanatory variables reinforces the nonparametric measure of efficiency by comparing the t statistics and the statistical significance of each estimated parameter.

The initial choice of potential variables must always be guided by the specifics of the problem. For example, one could reasonably postulate that research and development expenses affect productivity and sales, with a lag of several months or years. Similarly, some capital expenditures do not affect productivity immediately (new plant and equipment expenditures imply a construction lag or a training lag). The ultimate choice of the number of variables included in the analysis is also influenced by the number of observations available for the study. The larger the number of variables included, the larger the number of firms on the efficiency frontier, and the less valuable is the meaning of being an efficient firm or DMU (decision-making unit).

For these reasons, one output and six inputs were initially selected (the numbers in parentheses indicate the variable number for the Compustat database):

y_1 Total production; sales (12) plus changes in inventories (303)
  v_1 Interest expense (15)
v_2 Other production expenses; cost of goods sold (41) plus selling, general and administrative expense (189) minus interest (15), research and development (46) and advertising (45) expenses
v_3 Net physical capital; property, plant and equipment after depreciation (8)
v_4 Net investment (equal to capital expenditures (30) minus depreciation (14))
v_5 Research and development expense (46)
v_6 Advertising expense (45)

We initially regressed the six variables and their first lag on sales, using the R^2 forward sweep method for inclusion to identify the set of variables that best predict sales. All the possible subsets of variables were estimated for models with one variable, then two variables, and so on up to nine variables. The best models identified included the following variables:

	Number of dependent variables in the best model								
	1	2	3	4	5	6	7	8	9
v_1								Included	Included
v_2	Included	Included	Included	Included	Included	Included	Included	Included	Included
v_3				Included	Included	Included	Included	Included	Included
v_4							Included	Included	Included
v_5				Included	Included	Included	Included	Included	Included
v_6		Included	Included						
Lag v_1						Included	Included	Included	Included
Lag v_2									
Lag v_3						Included	Included	Included	Included
Lag v_4			Included	Included	Included				Included
Lag v_5									
Lag v_6					Included	Included	Included	Included	Included

Since the variables lag v_2 and lag v_5 were not selected in the forward sweep (based upon their contribution to improving the R^2 of the regression), they were removed from consideration for inclusion in the model.

After identification of the variables contributing the most to explain the behavior of the output, one must decide upon the number of input variables contributing the most to the measure of efficiency. The measures of efficiency were computed on a yearly basis, using an increasing number of inputs as identified in the previous step. As expected, the number of efficient DMUs increases with the number of inputs used in the analysis. Out of 112 observations, the numbers of efficient DMUs were 69, 80, 97, 103, 106 and 106 for analysis using 1 to 6 inputs respectively (see Tables 6.3 and 6.4). The addition of the sixth input did not change the number of efficient units. Since 95% of the units are already called 'efficient' with the use of six inputs, we did not consider the inclusion of additional inputs.

At the next step two regressions were run for each model to test the relationship between the chosen inputs and the output. The first regression used data for all the units, and the second used only data from the efficient units. The regression parameters, which estimate the relationship between one of the inputs and the output, $\beta_i = \partial y/\partial v_i$, can be viewed as a measure of the productivity of input i. If all inputs and outputs are expressed in dollars, the regression coefficient can be interpreted one plus as the average rate of return on investment in input i (e.g., a parameter estimate of 1.25 indicates that an expenditure of $1 on the input yields an average increase in revenue of $1.25). Hence if a parameter estimate is higher for the regression using only the subset of efficient

Table 6.3 Regression results with efficient DMUs

Number of regressors	**1**	**2**	**3**	**4**	**5**	**6**
Number of DMUs	69	80	97	103	106	106
Corrected R^2	0.9993	0.9996	0.9995	0.9995	0.9995	0.9996
v_2	1.17333 (<0.0001)	1.14772 (<0.0001)	1.13508 (<0.0001)	1.16678 (<0.0001)	1.17484 (<0.0001)	1.15758 (<0.0001)
v_3				−0.23630 (<0.0001)	−0.33309 (<0.0001)	−0.19599 (0.2020)
v_5				1.77320 (<0.0001)	2,87441 (<0.0001)	1.39081 (0.1037)
v_6		4.14306 (<0.0001)	3.98341 (<0.0001)			
Lag v_1						2.90533 (0.0022)
Lag v_3						−0.99474 (0.5898)
Lag v_4			0.42577 (0.0388)	0.48743 (0.0203)	0.46810 (0.0218)	
Lag v_6					−2.9246 (0.0452)	2.07200 (0.2409)

Table 6.4 Regression results with all DMUs

Number of regressors	1	2	3	4	5	6
Number of DMUs	112	112	112	112	112	112
Corrected R^2	0.9992	0.9994	0.9995	0.9995	0.9995	0.9995
v_2	1.17222	1.14641	1.13460	1.16711	1.17501	1.16718
	(<0.0001)	(<0.0001)	(<0.0001)	(<0.0001)	(<0.0001)	(<0.0001)
v_3				−0.23699	−0.33344	−0.36132
				(<0.0001)	(<0.0001)	(<0.0001)
v_5				1.77241	2.87371	3.32504
				(<0.0001)	(<0.0001)	(<0.0001)
v_6		3.91764	4.01923			
		(<0.0001)	(<0.0001)			
Lag v_1						2.18464
						(0.0223)
Lag v_3						−3.4789
						(0.0549)
Lag v_4			0.44542	0.48888	0.46865	
			(0.0200)	(0.0151)	(0.0180)	
Lag v_6					−2.9227	0.10129
					(0.0394)	(0.9541)

DMUs one can conclude that corresponding input influences the level of efficiency.

Six models were estimated twice; once for the whole set of DMUs, and once for the subset of efficient units. At each state, an additional variable was introduced according to the table above. Results for each of the models are presented in the next two tables where the entry for each variable is the parameter estimate, and the number in parenthesis is the level of significance. The initial variable included is v_2, which is the cost of production, excluding interest, research and development and advertising expenses. This is the primary variable related to the level of output, and enters every model identified with the initial regression. The parameter estimates are significant at the 1% level for both regressions. Since the estimate of the subset regression is greater than the estimate for the whole set of DMUs, one can conclude that v_2 is a relevant variable for the model. The second variable included is v_6, which is the cost of advertising. The parameter estimates for v_2 and v_6 are all significant at the 1% level for both regressions. Since the estimate of the subset regression is greater than the estimate for the whole set of DMUs, one can conclude that v_6 is also a relevant variable for the model. Inclusion of v_6 increased the corrected R^2 statistic from 0.993 to 0.996 for the subset regression.

Inclusion of the third variable, lag v_4, did not yield an improvement in the corrected R^2 statistic for the subset regression, and the statistic for the subset model is the same as for the whole model. Hence lag v_4 does not improve the measure of efficiency. This conclusion is further supported by the fact that the parameter estimate is less than one (an increase in net investment and the previous period increases sales by only a fraction of the increased investment). The inclusion of other variables yields no further increase in the corrected R^2 statistic, while producing unreliable parameter estimates.

Two broad conclusions may be derived from these estimates. One is the prominent role played by the R&D input v_5, where its marginal contribution $\partial y/\partial v_5$ exceeds 1.772 but the lag is not very important. Second, the net investment variable v_4 which equals capital expenditures minus depreciation turns out to be unimportant. The variable v_3 representing net physical capital has a negative impact, which may be due to the fact that the DEA model here considers only level efficiency rather than growth efficiency.

When we consider growth efficiency type DEA model in the form

$$\begin{aligned} &\text{Min} \quad \theta, \\ &\text{s.t.} \quad \sum_{j=1}^{N} \hat{c}_j \leq \theta \hat{c}_h, \quad \sum_j \hat{R}_j \lambda_j \leq \hat{R}_h \\ &\qquad \sum_j \lambda_j \geq 1, \quad \lambda_j \geq 0 \end{aligned}$$

We may obtain more directly the influence of growth of R&D as $\hat{R}_j$ on the growth of average cost $\hat{c}_j = \Delta c_j / c_j$, where c_j is average costs made up of all the costs above. Denoting by β_2 the dual variable associated with the R&D variable $\hat{R}_h$, its optimal value for the efficient firms turned out to be higher than one, for example, a selected sample is as follows.

	1988–91		1997–2000	
	$-\beta_2$	θ	$-\beta_2$	θ
Apple	1.26	0.90	1.21	0.98
Compaq	1.50	1.00	1.34	1.00
Dell	2.71	1.00	1.54	1.00
Sequent	0.80	0.72	0.95	0.94
Sun	0.83	0.76	1.79	0.97

Table 6.5 Regression results with three inputs for efficient firms

Period	Intercept	x_1	x_2	x_3	$\overline{R}^2$
1985–88	767.5	1.381*	0.487*	6.949**	0.829
1993–96	−146.6	−0.089	1.352**	2.544**	0.997
1997–2000	−239.9	−0.155	1.194**	4.001**	0.995
1985–2000	8.624	0.109	1.077**	4.294**	0.996

Note: One and two asterisks denote significant t values at 5% and 1% respectively.

It is clear that a high rate of R&D inputs growth tends to reduce the average cost rate for the efficient firms and this acts as a strong factor for entry deterrence. With a high price elasticity of demand, the market size effect operates over time and as discussed before it may indicate significant R&D externalities.

Finally, we can analyze the impact of R&D expenditures on output (net sales) by aggregating the six inputs into three groups: x_1 = net plant and equipment, x_2 = cost of goods sold and x_3 = R&D expenditures. The estimated regression results for the different subperiods are given in Table 6.5, which is based on only the efficient firms, which are on the production frontier. It is clear that the marginal contribution of R&D variable is significant and higher relative to the other two inputs.

6.4 Concluding remarks

Innovation efficiency through R&D has intensified the market dynamics and entry–exit behavior in the computer industry as measured by the changes in market shares. The leading firms who have remained dynamically efficient over time has exploited this R&D based innovation efficiency as a basis for cooperation. Recently Baumol (2002) has developed a technology-consortium model, where the cost of not joining the R&D or technology-consortium is very high for each member who does not join. This cost is due to lower profits and the expected opportunity loss. The advantage of joining a consortium is not simply due to cost reduction through scale economies and learning but also due to the fact that it discourages cheating in a game-theoretic sense. Furthermore the market share dynamics based on cost and price reduction in the computer industry has intensified this form of cooperative R&D behavior both at home and abroad, where the leading firms have exploited their efficiency to gain market share. Volatility in this market is expected as Metcalfe (1994), Mazzucato (2000) and Lu (2001) have shown that competition in modern research designs comes in many forms which foster cooperative and collusive behavior.

References

Bain, J.S. (1956): *Barriers to New Competition*. Harvard University Press, Cambridge.

Baumol, W.J. (2002): *The Free Market Innovation Machine: Analyzing the Growth Miracle of Capitalism*. Princeton University Press, Princeton.

d'Aspremont, C. and Jacquemin, A. (1988): Cooperative and noncooperative R&D in duopoly with spillovers. *American Economic Review* 78, 1133–7.

Demsetz, H. (1982): Barriers to Entry. *American Economics Review* 72, 47–57.

Fanchon, P. and Sengupta, J.K. (2001): The influence of lags on dynamic efficiency measures. Paper presented in Informs International Conference, Maui, Hawaii, June 17–20.

Geroski, P.A. (2003): *The Evolution of New Markets*. Oxford University Press, Oxford.

Gort, M. and Kleeper, S. (1982): Time paths in the diffusion on product innovations. *Economic Journal* 92, 630–53.

Gort, M. and Konakayama, A. (1982): A model of diffusion in the production of an innovation. *American Economic Review* 72, 1111–20.

Griliches, Z. (1957): Hybrid corn: an exploration in the economics of technological change. *Econometrica* 25, 501–22.

Hinloopen, J. (2003): R&D efficiency gains due to cooperation. *Journal of Economics* 80, 107–25.

Jovanovic, B. and Lach, S. (1989): Entry, exit and diffusion with learning by doing. *American Economic Review* 79, 690–9.

Kamien, M.I., Muller, E. and Zang, I. (1992): Research joint ventures and R&D cartels. *American Economic Review* 82, 1293–1306.

Lansbury, M. and Mayes, D. (1996): Entry, exit, ownership and the growth of productivity. In: Mayes, D. (ed.), *Sources of Productivity Growth*. Cambridge University Press, Cambridge.

Lu, D. (2001): Shared network investment. *Journal of Economics* 73, 299–312.

Mazzucato, M. (2000): *Firm Size, Innovation and Market*. Edward Elgar, Cheltenham, UK.

Metcalfe, J.S. (1994): Competition, evolution and the capital market. *Metroeconomica* 4, 127–54.

Norsworthy, J.R. and Jang, S.L. (1992): *Empirical Measurement and Analysis of Productivity and Technological Change*. North Holland, Amsterdam.

Philips, L. (1995): *Competition Policy: A Game Theoretic Perspective*. Cambridge University Press, Cambridge.

von Weizsacker, C.C. (1980): *Barriers to Entry*. Springer Verlag, Berlin.

Index